航空器适航技术系列教材

Airworthiness Design of Composite Structures

复合材料结构适航性设计

Li Longbiao   Tiniakov Dmytro

李龙彪   [乌克兰]蒂尼亚科夫·德米特里   主编

北京航空航天大学出版社

图书在版编目(CIP)数据

复合材料结构适航性设计 = Airworthiness Design
of Composite Structures：英文 / 李龙彪,(乌克兰)
蒂尼亚科夫・德米特里主编. -- 北京：北京航空航天大
学出版社,2021.8
　　ISBN 978-7-5124-3577-3

　　Ⅰ. ①复… Ⅱ. ①李… ②蒂… Ⅲ. ①民用飞机－复
合材料结构－适航性－结构设计－英文 Ⅳ. ①V257

中国版本图书馆 CIP 数据核字(2021)第 148423 号

Airworthiness Design of Composite Structures
复合材料结构适航性设计

Li Longbiao　　　　　Tiniakov Dmytro

李龙彪　　[乌克兰]蒂尼亚科夫・德米特里　主编
策划编辑 董 瑞　　责任编辑 江小珍

*

北京航空航天大学出版社出版发行

北京市海淀区学院路 37 号(邮编 100191)　http://www.buaapress.com.cn
发行部电话:(010)82317024　传真:(010)82328026
读者信箱: goodtextbook@126.com　邮购电话:(010)82316936
涿州市新华印刷有限公司印装　各地书店经销

*

开本:710×1 000　1/16　印张:14.25　字数:312 千字
2021 年 9 月第 1 版　2021 年 9 月第 1 次印刷　印数:1 000 册
ISBN 978-7-5124-3577-3　定价:46.00 元

Preface

Composite materials are composed of two or more kinds of materials with different properties. Through physical or chemical methods, these composite materials have new properties. The comprehensive properties of composite materials are better than those of raw materials, so as to meet different requirements. The history use of composite materials can be traced back to ancient times. Straw or wheat straw reinforced clay and reinforced concrete, which have been used for hundreds of years, are composed of two materials. In the mid-1960s, carbon fiber reinforced composites came into being, and they were used in aircraft structures in the early 1970s. Compared with the traditional materials, composite materials have the characteristics of high specific strength and specific modulus, good fatigue resistance, designability, and the identity of materials and structures. The application of composite materials in aircraft structure has been improved continuously. The amount of composite materials used in Airbus A350 is close to 40% of the total mass. The composite materials used in the wing and fuselage of Boeing 787 are more than 50%. Only carbon fiber composites used on fuselage panel for Airbus A380 is more than 30 tons.

The first concern of civil aviation is flight safety. Aviation administrations have been set up all over the world, and airworthiness regulations have been drawn up for civil aviation, and the safety requirements of civil aircraft have been scientifically and reasonably classified. Airworthiness refers to the quality of flight safety and physical integrity of civil aircraft (including the overall performance and control performance of its components and subsystems) under the expected service environment and use restrictions. This quality requires that the aircraft should always be in accordance with its model design standards and always in safe operation. This book introduces the theories and methods involved in the field of airworthiness design of composite structures for civil aircraft, including:

Chapter 1 is about general information about composite materials: history of their development, classification and composition.

Chapter 2 is about composite materials' components specifics: matrix and reinforcement performance for different composite materials type.

Chapter 3 is about main types of composite materials' production. There are descriptions about different production processes and their specifics.

Chapter 4 is about mechanical behavior of polymer composite materials.

Chapter 5 is about joints application in composite materials and their specifics.

Chapter 6 is about repair processes for composite materials parts according to their structural features.

Chapter 7 is about composite materials parts, units and products safety analyzing. There are descriptions about composite materials testing processes. Also, specifics of composite materials structures certification are given.

Li Longbiao, Tiniakov Dmytro

7 Dec. , 2020

Contents

Introduction

For modern engineering products, there are many strict and contradictory requirements, such as the structure's minimum mass, maximum strength, rigidity, reliability and durability to provide hard loading conditions and operations that include wide temperatures range and aggressive environments. The main class of materials, satisfying such stringent requirements, is composite materials.

The modern science of composite materials has dynamic development over the past decades, mainly because they are used in aviation and rocketry industry. The operating conditions of heavy loaded components and structural elements of the developed aircraft and rockets do not allow to use traditional structural materials. Each new project, which has much better performance, as a rule, requires the development of new structural materials. For example: as the project "Dreamliner" B787 of the Boeing Company. They wanted to decrease the airframe weight for the fuel consumption decreasing and passenger capacity and flight duration increasing. The B787 consists of modern composite materials about 60%. It is provided to decrease the fuel consumption about 20%, emission level about 25%, to increase strength of the fuselage and comfort conditions for passenger and crew (inner pressurization is equivalent 2 km altitude, for regular aircraft it is equivalent 2. 5 km), etc. Naturally, traditional structural metal materials cannot be used to make the structure with such a performance. An effective solution can be achieved only with the application of new high-strength, lightweight, rigid, high-temperature composite materials. Such materials are developed in laboratories of different countries.

The necessary to develop new structural materials is the basis for continuous improvement of manufacturing processes of their production, the fundamentally new technologies and equipment creation for their implementation. Within a relatively short time, new structural and manufactural solutions, developed together with new unique materials, are distributed in various industries. At the same time, new vehicles are being developed, the production and operation of which are impossible without new materials. Thus, the development of new structural materials, including composites, is one of the drivers for the engineering development.

In recent times, the composite materials production growth rate is constantly increasing. Twenty years ago, the total cost of composites produced only in the

United States exceeded $ 6 billion per year, and the total annual production of various types of composite materials was 3. 6 million tons.

Usually, the composite materials cost is very high, due to the manufactural processes complexity of their production and the high cost of the components which are used. However, reducing the joints' number, the parts' number and the assembly operations' number are the reason for the manufacturing costs decreasing. The composite materials' products production labor intensity can be reduced by 1. 5—2 times in comparison with metal counterparts.

Composite materials effectively compete with traditional structural materials such as aluminum, titanium, steel. The industries that actively use composite materials include aviation, astronautics, transport, chemical engineering, medicine, sports, tourism and education. Composites are used for the production of automobiles, railway transport facilities, airplanes, rockets, ships, submarines, containers for storing various kinds of liquids, pipelines, bicycles, sports equipment, etc. The special materials, the development of which was initially carried out by orders of the military departments, primarily for use in aircraft, have been applied in many sectors of the civil industry.

Modern composite materials' production has one important feature: the high demand for materials with new performance requires the rapid development of a large number of new materials. However, research on the properties of new composite materials is an expensive and time-consuming process. Moreover, such research often takes longer time than new materials' development. The designing process of products from composite materials has no less difficulties. Specialists need about 15—20 years for the perfect learning of the designing process of composite materials.

It is known that the structure's effectiveness is determined by the level of design solutions, the materials properties and the manufacture processes features of the structural elements from selected materials. In traditional engineering, examples of close interaction of designers, materials science engineers and manufacture engineers in time of creating any vehicles are quite rare. For modern composites materials' creation, this approach is unacceptable. The principle formulated in the early 1980s provides an integrated solution of choosing composite materials designing structures from them and developing manufacture processes. These are three sides of a single problem. Thus, in many cases, a new structure, a composite material and a producing process are simultaneously developed.

In modern time, all stages such as design, test, certification, production for civil aircraft and related system development have been carried out in parallel ways with accordance to technologies and industries levels. A civil aircraft depends on some engineering areas as shown in Figure I. 1. Civil aircraft is the end product of an assembly of systems, units, parts, elements and equipment.

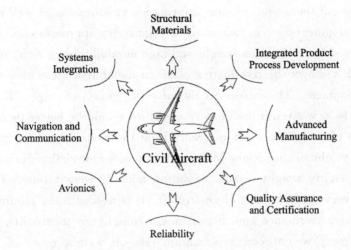

Figure I. 1   Commercial civil aircraft development layout

Civil aircraft are firstly needed to be designed to fully comply with requirements. From this point of view, structural material development (modern and future) should be estimated according to requirements of aircraft structure, structural integrity, fatigue and damage tolerance, maintainability, etc. Allowable stresses are determined by deep analysis on the base of strict tests. For the required durability, the airframe should demonstrate enough high fatigue endurance via its Design Service Goal.

New materials' application for civil aircraft industry is often interfered by the long-time and high cost of modern qualification procedures. A lot of money (maybe some million dollars) and some years are usually needed to complete new structural materials qualification process. The results of these researches depend on many factors and any changes in them can discommode this process. This fact is a serious problem for new structural materials implementation. The integral materials engineering can decrease this problem. The development of a material performance prediction according to processing and critical specifics of their application can improve and optimize their qualification.

Providing a civil aircraft project to engineering and commercial successes is not a simple task. Modern structural material availability is important, but their successful and reasonably priced application is engineering task. Integrated Product and Process Design and Concurrent Engineering approaches are developed and are widely being applied in the last years. Also, Product Life Cycle Management tools, software, process, etc. are used for these purposes. All of these things help researchers to certify commercial civil aircrafts successfully. Product Life Cycle Management provides design, engineering analysis, technical documentation and the integration of the overall product information environment as well as named as Product Development System. All of these engineering approaches and tools do not automatically ensure minimum weight and high durability of an airframe.

Figure I.2 shows the risk matrix of seven major risk areas of a civil aircraft project development. These risks are linked with the project stage. "Right Model" means that the new aircraft preliminary structure should be better than its nearest analogs both technically and commercially from the beginning of the project. One of the tasks to obtain the "Right Model" in time of a new civil aircraft preliminary stage is the empty weight versus maximum take-off weight ratio. That is also related to Structural Efficiency of an aircraft. It is needed to be minimized within the structural performance and airworthiness compliance constraints. Table I.1 shows the empty weight versus maximum take-off weight ratios of some well-known civil aircraft.

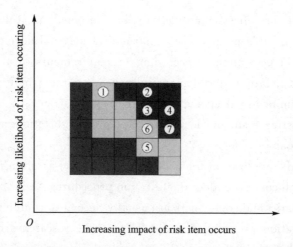

1—"Right Model"; 2—tech for design; 3—skilled staff; 4—design org;
5—vendor supply chain management; 6—production org; 7—certification

Figure I.2   Major risk areas of a civil aircraft project development

Table I.1   The empty weight versus maximum take-off weight ratios off some civil aircraft

Selected models	"Structural Efficiency" ratios of empty weight to maximum take-off weight
Concorde	0.425
Tu – 144	0.471

(**continued**)

Selected Models	"Structural Efficiency" ratios of empty weight to maximum take-off weight
B777 – 200LR	0. 426
A340 – 500	0. 460
IL – 62M	0. 434
B777 – 300ER	0. 473
B787 – 8	0. 482
B787 – 9	0. 464
B767 – 300ER	0. 482
IL – 96MT	0. 491
B747 – 8	0. 482
MC – 21 – 300	0. 458
SSJ – 100	0. 491

Structural Efficiency on the one side, is directly based on the fuel consumption, cost of fuel and also, at the modern time, with released carbon-dioxide emission. On the other side, the empty weight reducing is an inversely influence on the airframe strength. The aircraft must comply with the strength and service life durability requirements as per Airworthiness. These two opposite tasks create the structural material selection, design and manufacturing ways.

It is forecasted that fuel efficiency and operational productivity of the next generation of civil aircraft should be improved between 60%—70% by 2030—2040.

This can be obtained primarily by two approaches: by the overall fuel consumption decreasing and by the engine's efficiency increasing. For these aims, the structure part at the take-off weight should be decreased by 5%, while the total empty weight should be decreased about 30%. In addition, the powerplant part at the take-off weight should be decreased by 3%. In order to achieve these aims, new structural materials will be needed for the next generation of civil aircraft.

The applications of composite materials for new structures are expected to be increased in the next generation of civil aircraft.

The applications of composite materials have been continuously increasing since the 1990s as shown in Figure I. 3 and new Boeing 787 "Dreamliner" and Airbus 350 XWB series aircraft are utilizing the highest amount of composite materials.

The A350 XWB's airframe materials were chosen for their rational

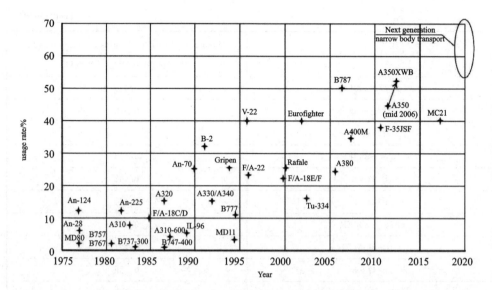

Figure I. 3　Chronology of composite material usage in aircraft

performance. Composites are in the fuselage, wings and tail. Advanced metallic alloys are in the landing gear, engine and beams. "Intelligent airframe" is new Airbus' philosophy. It is about the rational material application for each separate case. Airbus uses composite materials widely —involving 53% of the full airframe (compared to 11% in the A330). This way provides profits from such lightweight, strong and durable materials. Their advantages are decreased development times, high production rates on the final assembly stage, low total aircraft weight, high service durability, improved corrosion and fatigue stability, as well as low maintenance costs. The composites' application in the fuselage, wing and tail is supposed to reduce maintenance of operations by creating the "intelligent airframe" with improved corrosion and fatigue resistance during the lifetime.

Boeing Company in 2004 proposed the B787 "Dreamliner", an all-new airplane with the highest efficiency. The airplane fuel consumption is 20% less than modern similar civil airplanes. The key-point to the B787 high performance is its set of new technologies and structure. Composite materials make up 50% of the B787 airframe.

The applications of composites in engines are also expanding. The main effort in this direction is the Advanced High Temperature Engine Materials Technologies development that is based on high-temperature composite materials creation: up to 425 ℃ for polymer-matrix composites, up to 1, 250 ℃ for metal-matrix/inter-metallic-matrix composites, up to 1, 650 ℃ for ceramic-matrix composites.

Based on the preliminary researches for the next generation of aviation engines, material temperatures incoming 1,650 ℃ are expected for the turbine inlet. It requires wide ceramic composites application in the combustor, turbine, and nozzle. One of the advantages of ceramic composites application is that they provide higher operational temperatures and higher combustion efficiency with decreased fuel consumption. The low density ceramic composites provides a 50% reduction in the engine weight. Ceramic matrix composite for turbine blades are useful due to their high temperature resistance. Without the necessity to cool the turbine the higher temperatures can be provided in the combustor. The development of new ceramic composites is aimed at obtaining reliable ceramic composites reinforced by continuous ceramic fibers like regular monolithic ceramics, such fiber-reinforced ceramics have lower densities, higher oxidation resistance and higher operational temperatures than traditional alloys. In addition, unlike monolithic ceramics, fiber-reinforced ceramics show metal-like deformation performance, non-critical damages, and strength that have low sensitive to processing- and service-generated mistakes.

Polymer composites are the lightest of composite materials and their applications in aircraft powerplant, such as General Electric Company engines, have led to reduction in both engine weight and production costs. However, to realize more advantages of polymer composites in aircraft powerplant, new composite materials should be developed with higher thermal and oxidation resistance that permit their application at operational temperatures up to 425 ℃.

In the civil aircraft industry, development, design, production, operation, flight and ground staff training, maintenance, air traffic control and all linked processes are regulated by international rules and organizations. The highest regulatory organization is the International Civil Aviation Organization (ICAO) in 1944. Under ICAO regulations several National Civil Aviation Authorities were created such as EASA (European Aviation Safety Agency) in Europe, FAA (Federal Aviation Authority) in the United States. China operates and maintains aircraft according to Civil Aviation Administration of China (CAAC), which are fully compliant with EASA and FAA regulations. The full aircraft life cycle, from design stage to operation must be certified by the authorized organization.

Aviation products, their application and organization approval (aircrafts, engines), top down regulation structure of EASA is given in Figure I. 4. Major regulations for civil aircraft airworthiness are defined by Annex 8 of the Chicago Convention. EASA CS25, FAA FAR25, CAAC CCAR25 regulations define the design and certification requirements for transport category aircraft.

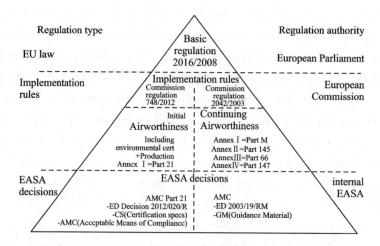

Figure I. 4   Top Down Structure of EASA Civil Aviation Regulations

EASA Certification of Product, Parts and Appliances:

1) Product

① Design Organization Approval (Subpart J);

② Product Organization Approval (Subpart G).

2) Aircraft (Type Certification) Certification Basis for Large Aircraft (CS25).

3) Engine (Type Certification) Certification Basis (CS-E).

4) Propeller (Type Certification) Certification Basis (CS-P).

5) Change to Type Certifications:

① Design Organization Approval (Subpart j);

② Production Organization Approval (Subpart G);

③ STC (Supplemental Type Certification);

④ Major Changes/Minor Changes.

6) Parts and Appliances.

7) ETSO (European Technical Standard Orders) Parts.

The principle of continuing Airworthiness is tightly connected with an aircraft and its systems reliability. Some systems and units can operate without maintenance of any type. For bigger part systems there are two kinds of maintenance. One or both of them can be applied. In preventive or scheduled maintenance, elements and parts are replaced, and oils are changed, or regulations made before failure take place. The target is to increase the system and unit's reliability above the term for preventing the aging effects, wear, corrosion fatigue, and linked events. While, adjusted or unscheduled maintenance is carried out after failure has taken place with aim to turn the unit or system to service conditions as soon as possible. Such maintenance operations are carried out at unpredictable terms as the time to any certain unit's failure cannot be set ahead of time.

In general, structural design load and damage approval of Airworthiness requirements (CS25, FAR25, CCAR25) define Limit Load and Ultimate Load. The Limit Load is defined as being the maximum load per life which may only cause a detectable damage to be found and repaired through maintenance. Ultimate Load is 1.5 times of Limit Load and it is allowed to cause only an acceptable but non-detectable damage which is referred as the Allowable Damage Limit.

The major Fatigue and Damage Tolerance criteria from which the airframe is designed and certified, is to guarantee compliance with the airworthiness, with taking into account durability, interval and threshold inspection, and repairability of the given structure. Allowable stresses are determined by analysis on the base of the given series tests. The allowable stresses are dependent on the structure shape and dimensions, the applied material and loading configuration. In time of an aircraft structure designing and engineering analysis, Reserve Factors or Margins of Safety are determined.

Margins of Safety = Allowable stresses/Calculated stresses−1   (I.1)

About the durability, the airframe should be designed to show enough high fatigue endurance via its Design Service Goal, to reach the following targets:

1) Provide the structure durability via its operational life.

2) Minimize the number of zones inclined to fatigue failure and cracks development in service.

Structural analysis, geometric determination and tests must show that the calculated fatigue life of the structure is higher or equal to the design service goal that multiplied by an appropriate scatter factor.

Fatigue life ⩾ Design service goal × Scatter factor   (I.2)

The Scatter Factor value depends on the stress design data applied in the analysis.

The limit for the structure initial inspection should be determined as a design target. Design precautions should be taken for the following targets:

1) Provide the minimum inspection limit should be equal to the target value.

2) Provided that any damage will not obtain its critical value before the first inspection.

Tests and analysis should show that the aircraft service life is higher or equal to the design target inspection limit multiplied by a corresponding scatter factor.

Service life⩾ Design target inspection limit × Scatter factor.   (I.3)

The inspection limit evaluation, applying initial defect concept, must guarantee that cracks will not extend from the initial defect to the critical value within the inspection limit interval. This way is appropriate to Single Load path structures and Multiple Load Path structure where it cannot be shown that load path failure, partial failure, or crack retardation can be detected and repaired along regular maintenance.

The repeat inspection interval is the time between two subsequent directed inspections along that any damage must not extend from the detectable value to the critical value. Design precautions should be taken for the following targets:

1) Provide structural damage resistance capability of the structure.

2) Provided that any failure can be determined before it reaches critical value within the targeted inspection interval.

3) Maintain airworthiness by scheduled inspections.

Repeat inspection interval is determined from the time period which failure is detectable, and the permanent strength stays higher than required level. Therefore, the structural evaluation should include a determination of the period of crack to develop from the detectable value to the critical value under permanent loads in the critical location of the structure. Then an inspection interval is determined by using a corresponding scatter factor to this crack growth time, in order to guarantee that the crack can be detected before the residual strength of the structure is reached.

In determining the inspection interval, the detectable crack value should be consistent with the specifics of the proposed inspection approach. The crack analysis and crack growth test must show that the period along which the crack grows from the detectable value to the critical value, is higher and equal to the required repeated inspection interval, multiplied by a corresponding scatter factor.

Period of the crack grows $\geq$ *Inspection interval* $\times$ *Scatter factor.*   (I. 4)

Frequent buckling should be escaped because it has an effect on the skin and webs fatigue lives and its nearby structures. The fatigue load spectrum should be analyzed to provide that the buckling will not occur more than given number along the aircraft service life. The allowed numbers repeated buckling should be proved by tests under fatigue damage to structure buckling within the fatigue load range.

Repairability is a performance of the structure that related to the structure ability to a reasonable repair with the minimum of structural changes following the occurrence of damage. Repairability should be considered in the aircraft design. Repairability is improved with improving of accessibility, serviceability and unification while corrosion are minimized. Moreover, the maximum application of interchangeable elements becomes suitable that can support fast repair and replacement. Reparability of an idea, limit and ultimate loads capability of an aircraft structure is shown in Figure I. 5.

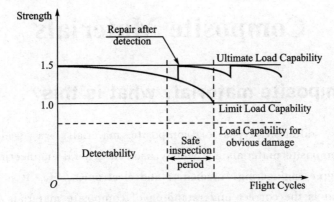

Figure I. 5 Aircraft structures repairability iilustration

This book is useful for students majoring in different aviation or material sciences and experts in related engineering areas.

Chapter 1 General Information About Composite Materials

1.1 Composite material, what is this?

Currently, different types of composite materials are used in various industries. Composite materials are widely used in general engineering, aviation, rocket and engine engineering, medicine, and electronics, etc. It is necessary to determine what is the correct understanding of composite materials by engineers from various fields of science and technology.

The word "composition" is widely used in art. It comes from the Latin word "*compositio*" which means compilation, linking. Proceeding from this, there is, for example, an art composition or a musical composition, which consists of various parts and passages. If we use these considerations to the concept of "composite material", then in the most general case, it can be noted that composite material (CM) is a material consisting of (composed) some different parts.

Sometimes CMs include structural materials with protective coatings. For example, Dominic V. Rosato thinks that CMs are elements coated with polyvinyl film used in aircraft, laminated metal-plastic lining and other elements of this kind.

There is a definition according to which CMs are materials that consist of two or more components (reinforcing elements and a matrix holding them together) and have performance different from the components performance. It is assumed that the components that make up the CM should be well compatible and not dissolve or otherwise absorb each other.

The interpretation width of the word "composite" has been given to many definitions of the term "composite material". In a broad sense, a CM is any material with a heterogeneous structure, i. e. with a structure consisting of at least two phases. Such a definition makes it possible to classify the big part of modern metallic materials as CMs. A typical example of a heterophasic metallic material is steel with a plate-perlite structure, which is a mechanical mixture of ferrite and cementite plates. Given the above definition, polymeric materials can also be attributed to CMs since in addition to the main component (resin), they contain various fillers, dyes, etc.

If we analyze the natural materials structure, then they can also be attributed to CM. For example, wood is a composition of cellulose fibers, held together by a matrix of an organic substance —lignin.

For the first time, CM began to be created long before the concept of "composite material". For example, bricks made from clay and straw reinforced.

The first examples of a scientific approach to the artificial CM's creation are the development of armored concrete and fiberglass. Armored concrete is one of the first samples of reinforced ceramics. This composite consists of concrete and steel reinforcement. Concrete perfectly perceives compressive loads and very poorly resists tensile stresses. For this reason, the ceilings of buildings cannot be made of ordinary concrete. When concrete and rods steel are combined, which are located in a certain way and are well supporting tensile loads, armored concrete is obtained the advantages of both components.

The first patent for composite polymer material (synthetic resin reinforced with natural fibers) was issued in 1909. But, it should be noted that the ancient Egyptians built river boats from reeds soaked in a resin.

There were some attempts to give a clearer scientific definition of artificial composite materials in order to mark their specific features. However, it is noted that at present there is no generally accepted definition of artificial CM. The most completed is the definition according to which CMs include materials with a number of features:

1) the material components composition, shape and arrangement are "determined in advance";

2) the material does not occur in nature, but is created by man;

3) the material consists of two or more components that differ in chemical composition and divided by obvious boundary;

4) the material properties are determined by each of its components, which should be present in sufficiently large quantities;

5) the material has such properties that its components, taken separately, do not have;

6) the material is heterogeneous on a micro-scale and homogeneous on a macro-scale.

According to opinion of some specialists, the sixth feature does not allow bimetals and coated materials to be classified as composite materials, since they are not homogeneous on a macro-scale. These materials are not characterized by repeating geometry or uniform distribution of components with respect to each other.

Thus, composite materials are characterized by features noted above, according to which they should be especially separated from a wide range of structurally heterogeneous materials. Graphically, the place of composites in the group of widely used structural metal materials with structural heterogeneity is shown in Figure 1.1.

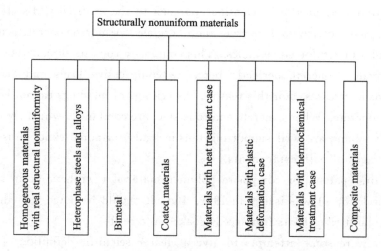

Figure 1.1   Types of materials with structural nonuniformity

1.2   Composite material classification

Currently, the branches of science and production related to the development of new CMs are developing most dynamically. Specialists have developed and researched a large number of CM. Analysis of the CM production development prospects shows that in the nearest future the range of these materials will rapidly expand. In order to systematize CM according to various criteria, correct implementation of the CM choosing procedure for the manufacture of various elements, a rational classification of these materials and is necessary.

There is no single generally accepted CM's classification. There are several reasons for this. One of them is that CMs represent the widest class of materials, combining metals, polymers and ceramics. Modern technologies allow the various implementation of materials combinations. The variety of base materials and the real structures complexity complicate the classification of resulting CM. In addition, the term definition "composite material" is difficult.

Currently, several approaches to the CM classification have been developed. The CM classification can be carried out on a number of features.

The most commonly used CM classification is based on material attribute. In Figure 1.2 we can see one of the simplest CM classifications according to the type of matrix material. In accordance with this classification, CM are divided into polymeric CM (PCM), metallic CM (MCM), ceramic CM (CCM), carbon-carbon CM (CCCM) and hybrid CM (HCM). Hybrid composites are mixed matrix materials.

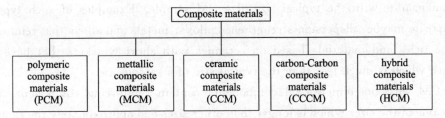

Figure 1. 2   Common composite materials classification

CM classification can be classified by the following features:

1) geometry of CM components;

2) components space arrangement (reinforcing scheme);

3) nature of CM components.

1.2.1   Composite material classification by components geometry

There are three groups of CM (Figure 1. 3):

1) Zero-dimensional components;

2) One-dimensional components;

3) Dimeric components.

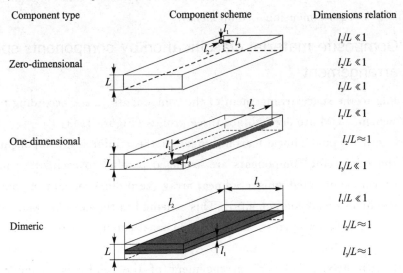

Figure 1. 3   CM classification by geometry of their components

This classification of CM is based on the concept of an elementary sample of CM, i. e. , such a minimal amount of a material has all the set basic features of

CM. In principle, an elementary sample should have all dimensions exceeding the minimum component size.

For the zero-dimensional CM, all three components dimensions are the same magnitude. The zero-dimensional components do not have a dimension that is commensurate with the typical size of a CM sample. Examples of such type of composite maybe: dispersion-strengthened alloys, metals and alloys that reinforced by particles and materials based on ceramics with short whiskers that have the length which is much less than the typical size of a sample.

CM with one-dimensional components contain reinforcing components that have one of the sizes which is longer than other sizes and approximately the same as a CM sample dimension. Examples of such CM: fibrous composite materials based on metals and polymers, which reinforced by ceramic, boron, carbon, glass fibers.

CM with dimeric components contain reinforcing components that have two sizes which exceeding the third size and commensurate with the typical size of a CM sample. Examples of such CM types are layered CM that consist of alternating layers of titanium and aluminum and their alloys.

Zero-dimensional, one-dimensional, and dimeric components have the next indices: 0_*, 1, and 2.

In addition to such CM, there are combined CM that have some components with various dimensions. An example of such CM is plastic based on epoxy resin, which reinforced by carbon fiber (one-dimensional component) and short nitrogen carbide whiskers (zero-dimensional component).

1.2.2　Composite material classification by components space arrangement

According to the space arrangement of the components, i. e. , according to the reinforcing scheme, CM are divided into three groups (Figure 1.4):

1) CM with uniaxial (linear) arrangement of the reinforcing component. In such CM, the reinforcing components are arranged parallel to each other in the matrix. They can be oriented whiskers linear array (zero-dimensional components) or fibers (one-dimensional components). This scheme has the next marking: 0_* :0 :0 and 1:0:0 (zero-dimensional 0_* or one-dimensional component is located along the x axis).

2) CM with biaxial (planar) arrangement of the reinforcing component. Reinforcing components are located in the matrix in planes parallel to each other. Such a reinforcing scheme is created by zero-dimensional (whisker mats), one-dimensional (fibers), or dimeric components (foils) and is marked by 0_* :0_* :0,

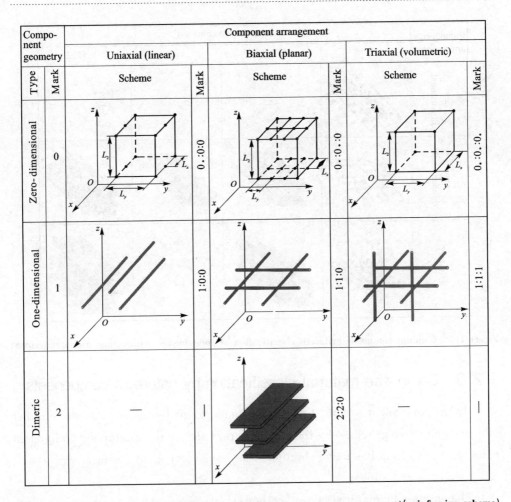

Figure 1.4　Composite materials classification by components space arrangement(reinforcing scheme)

$1:1:0$ and $2:2:0$, respectively (the components are located in the xy planes).

　　3) CM with a triaxial (volumetric) arrangement of components. As for this reinforcing scheme, it is impossible to select one or two preferred directions in the material. Such a scheme can be created by using zero-dimensional $(0_*:0_*:0_*)$ or one-dimensional $(1:1:1)$ components.

　　Figure 1.4 shows schemes for reinforcing CM with components of the same dimension (zero-dimensional, one-dimensional, or dimeric). For combined CM, i. e. materials reinforced with components of different dimensions, the following combinations of components are possible: 0_*+1; 0_*+2; $1+2$; 0_*+1+2. The most complex variants for reinforced combined CM are shown in Figure 1.5. A three-axis reinforcing scheme with a combination of zero-dimensional and one-dimensional components is shown by the way $(0_*+1):(0_*+1):(0_*+1)$. Also, two more variants are possible $(0_*+1):0_*:0_*$ and $(0_*+1):(0_*+1):0_*$.

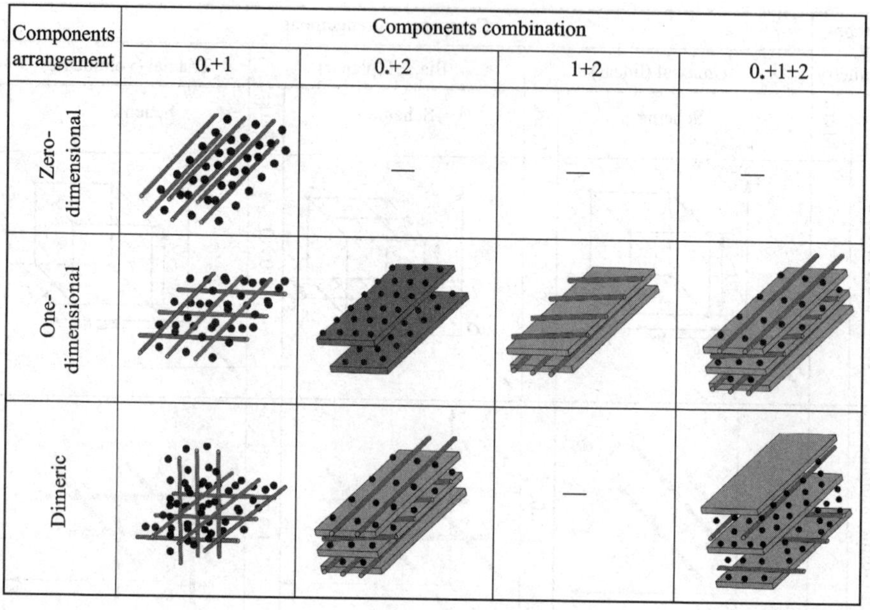

Figure 1.5　Combine composite materials classification by components combinations and arrangement

1.2.3　Composite material classification by nature of components

By the component nature, CM can be some groups by the components number (for example, two groups —by the matrix nature and by the reinforcing component nature). Each group has four subgroups that have components from:

　　1) metals and alloys;

　　2) non-metallic materials (e. g. carbon);

　　3) inorganic compounds (oxides, carbides, nitrides, etc.);

　　4) organic compounds.

In accordance with this classification, the carbon-carbon CM are CM with a non-metallic matrix and CM with non-metallic reinforcing component; the carbon CM are CM with a matrix of organic compound and CM with the reinforcing component from non-metallic elements.

Additionally, for this classification, there are poly-matrix CM and poly-reinforced CM. The poly-matrix CM consist of alternating layers of two or more different chemical composition matrices [Figure 1.6 (a)]. The poly-reinforced CM contain two or more different composition reinforcing components that uniformly distributed in the matrix [Figure 1.6 (b)].

Poly-reinforced CM can be simple if they contain reinforcing components of different nature, but because of the similar dimension, and combined if they contain reinforcing components of different dimensions and different nature. For example, fiberglass is a simple CM, and boron-aluminum composite with titanium foil interlayers is a combined poly-reinforced CM.

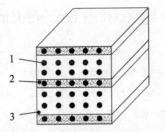
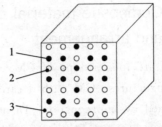

(a) In a poly-matrix CM there are two types of matrix (or more than 2)

(b) In a poly-reinforced CM there are two types of fibers (or more than 2)

1—fiber 1; 2— fiber 2; 3—matrix 1; 4—matrix 2

Figure 1. 6   CM structure

1.2.4   Composite material classification by matrix material

This classification is based on material science principle. In accordance with, one of the most important features of the CM classification is the matrix material and reinforcing elements. The general name CM, as a rule, comes from the matrix material. CM with a metal matrix are called metal composite materials (MCM). CM with a polymer matrix are called polymer composite materials (PCM), and inorganic-inorganic composite materials. CM containing two or more matrix materials of different composition are called poly-matrix.

For CMC description, a double designation is often used: first is matrix material, then fiber material. For example, the description copper-tungsten (Cu-W) corresponds to a composite material in which the matrix is copper and the fibers are tungsten.

Inorganic CM are described by the same way as MCM, i. e. in the double name, the first position refers to the matrix, and the second to fiber. For example, the description of alumina-molybdenum (Al_2O_3-Mo) corresponds to a CM with an aluminum matrix and molybdenum fibers.

1.2.5   Composite material classification by reinforcing
components geometry

In accordance with the reinforcing components geometry (powders or granules, fibers, plates), CM are divided into powder (granular), fibrous and lamellar. The first group includes dispersion-strengthened CM. The second group includes CM reinforced with continuous and discrete fibers, for example, an aluminum-boron fiber composition. The third group includes CM reinforced with continuous and discrete plates, for example, layered CM consisting of alternating foils of steel, aluminum and titanium.

1.2.6 Composite material classification by components structure and arrangement

There are four groups of CM. They are a carcass, matrix, sandwich and combined structures. CM with a carcass structure include, for example, pseudo alloy obtained by impregnation. Dispersion-strengthened and reinforced CM have the matrix structure. Materials with a layered structure include compositions obtained from a set of alternating foils or sheets of various nature or composition materials. The combined materials contain combinations of the first three groups. For example, pseudo alloys, which has carcass strengthened by dispersed inclusions, belong to CM, combining the carcass and matrix structures.

1.2.7 Composite material classification by reinforcing scheme

This classification is based on designing approach. There are two groups — isotropic and anisotropic. Isotropic materials are those that have the same performance (e. g. , strength) in all directions. The performance of anisotropic materials depends on the direction in the designed object. Anisotropy of CM is designed: it is specially applied when the material develops in order to provide the necessary object performance.

CM with a matrix structure can be randomly reinforced and orderly reinforced (Figure 1. 7). Randomly reinforced CM contain reinforcing components like

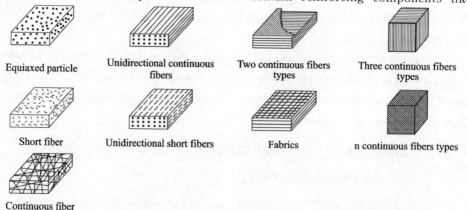

(a) randomly reinforced CM   (b) orderly reinforced CM   (c) orderly reinforced CM   (d) orderly reinforced CM
(uniaxial and space-network reinforcing types)

Figure 1. 7   CM classification by reinforcing scheme

dispersed inclusions, discrete or continuous fibers. These materials are isotropic or quasi-isotropic. The term "quasi-isotropic" means that the CM is anisotropic in micro volume, but isotropic in the total object. Orderly-reinforced CM can be unidirectional, i. e. , uniaxial-reinforced, biaxial-reinforced (with a planar arrangement of reinforcement) and triaxial-reinforced (with a volumetric arrangement of reinforcement).

1.2.8 Composite material classification by production technology

This classification is based on manufactural approach. In accordance with this classification, CM are produced by liquid-phase methods and solid-phase methods, as well as by deposition and spraying methods and combined methods.

Liquid-phase methods include impregnation (impregnation of reinforcement with polymers or molten metals) and directional crystallization of alloys. Solid-phase methods include rolling, extrusion, forging, stamping, explosive ramming, diffusion welding, drawing, etc. CM produced by solid-phase methods are used in the form of powder or thin sheets. CM that consist of the set of alternating matrix layers in the form of thin sheets (foils) and reinforcing components are called sandwich.

CM productions by deposition and spraying methods are as follows: the matrix is overlayed to the fibers from saline or other compounds, from the vapor-gas phase, from plasma, etc. Combined methods consist of several methods in the sequential or parallel application. For example, plasma spraying can be used as a preliminary manufactural operation, and rolling or diffusion welding can be used as the final operation.

1.2.9 Composite material classification by application

This classification is based on application approach. This classification is rather arbitrary, because composites are multifunctional materials. Nevertheless, CM can be: general structural materials (load-carrying structural member of ships, aircraft, automobiles, etc.), fire-resistant materials (aircraft turbine blades or combustion chambers, etc.), heat-resistant materials (elements operating in conditions of frequent heat exchange), friction materials (brake pads), anti-friction materials (plain bearings), impact-resistant materials (armor of aircraft, tanks), materials with special properties (magnetic, electrical, etc.).

1.3 Composite materials in the aircraft industry

The aviation industry is one of the main customers and consumers of CM. Figure 1.8 presents the specific part of CM in the weight of such craft as combat airplanes and helicopters, transport and passenger airplanes. The CM application in the aircraft structure allows to reduce their take-off weight, increase their payload, speed and range.

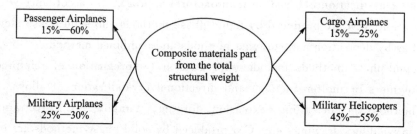

Figure 1.8   Composite materials application in the aviation industry

The aviation industry is widely used and continue to use various types of CM, including polymeric, metallic, ceramic, carbon-carbon, and hybrid. In 1958, the Boeing company applied fiberglass to manufacture a DC–8. Materials of this type are used for the manufacture of elements of sports aircraft too. In modern time, Boeing Company applied carbon fiber reinforced polymer (Figure 1.9) for its "Dreamliner" B787.

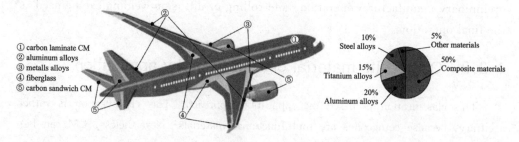

(a) Structural materials distribution at the airframe   (b) General distribution diagram of structural materials

Figure 1.9   Materials distribution by the mass in the B787 structure

One of the most widely used types of CM in aircraft structure are carbon-carbon composites. In Figure 1.10 the materials content in an airframe at different times are shown. At the end of the 20th century, the main materials for the airframe were aluminum alloys (about 80%). Due to the tightening of requirements over some decades, the volume of their application decreased by a substantial margin (about 11% at the beginning of the 21th century). The share of carbon plastics increased from 3% to 65%.

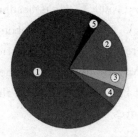 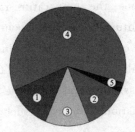

(a) at the end of the 20th century　　　　　(b) at the beginning of the 21st century

1 —aluminum alloys, 80%→11%; 2 —steel alloys, 14%→11%; 3 —titanium alloys, 2%→12%;

4 —Composite materials, 3%→65%; 5 —other materials about 1%

Figure 1.10　Structural materials distribution in an airframe at different times

The airframe of the experimental aircraft Voyager (Rutan Aircraft, USA), flied around the world in 1986 without intermediate landings and refueling, was totally made of carbon fiber.

The technical and economic efficiency of carbon fiber application in the heavy cargo aircraft Antonov An-124 airframe is shown in Table 1.1. 200 various parts with a total weight of 2.2 tons are made from carbon fiber of the KMU-3 type. The replacement of aluminum alloys due to the application of carbon fiber provides total weight decreasing about 6 tons and fuel consumption about 1.2×10^4 tons.

Table 1.1　The technical and economic efficiency of carbon fiber application in the heavy cargo aircraft Antonov An-124 airframe

Weight of the composite material parts/t	2.2
Numbers of parts	200
Total weight decreasing/t	0.8
Number parts decreasing	1,200
Total labor intensity decreasing/%	180
Aluminum alloys save/t	6
Freight traffic increasing/t · km	1×10^6 t · km
Fuel consumption decreasing/t	1.2×10^4

The modern CM using can be analyzed as an example by applying them into the gas-turbine engines structures. Engine's parts operate in very hard conditions: high temperature, vibrations, high level mechanical stress. They must well resist creep, withstand frequent heat exchanges, be lightweight, heat resistant, etc.

The key factor determining the gas-turbine engine efficiency is its operational temperature. Engine performance increase substantially with an operational temperature increasing. The temperature distribution in aircraft gas-turbine engine various zones is shown in Figure 1. 11. In the combustion chambers of modern aircraft engines, the temperature is about 1,300—1,500 ℃. It is assumed that in the nearest future it can reach about 1,800—2,000 ℃. A temperature about 2,000 ℃ is considered optimal for the engine, since it represents a stoichiometric limit for the ideal combustion process of all types of fuel. Any applied metal or modern alloy cannot successfully be used for the manufacture of heavily loaded parts that operate in the combustion chamber at such high temperatures.

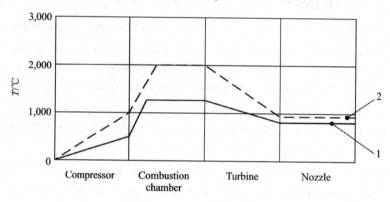

1—modern engines; 2—future engines

Figure 1. 11 Operational temperatures distribution into an aircraft gas-turbine engine

The most important part of a gas-turbine aircraft engine is a turbine blade that operates at high temperatures and a high level of mechanical stress. In the beginning, turbine blades were made by forging. The next step was the production of directionally crystallized blades with a columnar crystalline structure. A further increasing of the operational temperatures and engine performance was made possible by using single-crystal blades with internal cooling channels. Compared with directionally crystallized blades, the service life of single-crystal blades is 2—3 times longer.

The development and applying of CM with a ceramic matrix are one of the possible solutions to increase the working temperature for designing modern engines (Figure 1. 12). When the engine's maximum operational temperature increases up to 1,500 ℃, its efficiency will improve and it will use cheaper fuels. The disadvantage of ceramic matrix materials is the low level of reliability. The critical defect size for the ceramic materials is less than 100 μm, but for the metal materials, the presence of such defects up to some millimeters size is possible. A further operational temperature increasing is possible only with the development of new, more advanced structural materials.

Over the past decades, the CM application has steadily increased for the high-

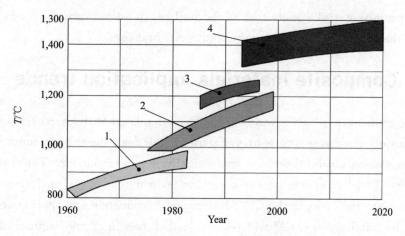

1 —traditional heat resistance alloys; 2 —directionally crystallized alloys;
3 —thermal resistance cover; 4 —CM with ceramic matrix

Figure 1.12   Application analysis of the heat resistance materials in aircraft engineering industry

temperature channels manufacturing of aircraft engines (Figure 1.13). At the same time, for the same engine parts, the relative volume of titanium alloys, steels, and other heat-resistant alloys is gradually decreased.

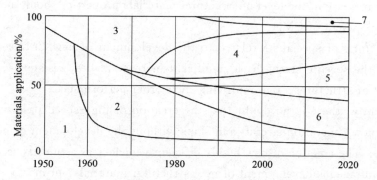

1 —steels; 2 —titanium alloys; 3 —nickel heat-resistant alloys; 4 —powder alloys;
5 —directionally crystallized alloys; 6 —composite materials; 7 —ceramic

Figure 1.13   The structural material application in the aircraft engines high-temperature paths in the second half of the 20th century

The aviation and aerospace industry are the main consumer of boron-fiber reinforced plastic. This composites type has high effectiveness for the structures that operate under compressive stresses and require high rigidity. Boron-fiber plastics can be used in combination with carbon-fiber plastics. The part of the structure that undertakes compressive stresses is made of boron plastics, and the part that undertakes tensile stresses is made of carbon plastics. With the same strength, the mass of beams of this type is 20%—30% lower than the mass of beams made of traditional aluminum alloys. The boron plastics application in the landing gear parts, fuselage sections, skin, gas-turbine engines compressor disks, etc. is promising. The high cost of boron-fiber plastics (mainly boron fiber) inhibits their wide application in other industrial branches.

1.4　Composite materials application trends

The challenges of CM's wide application are related with the development of new types of reinforcing and matrix materials, the development and improvement of existing manufactural processes, and a significant cost reduction. The analysis of these three listed directions shows a very wide field that opens up for specialists that related to the tasks of the CM development, production and application.

In the coming years, CM will retain a leading role in engine engineering, ship and aircraft engineering, and some areas of electrical engineering. The possibilities of a significant improvement in traditional structural materials properties through the using of existing approaches based on optimization and changes in the modes of their thermal treatment are close to exhaustion. An analysis of different solutions implemented in materials science over the past decades indicates that the most significant increase in the level of structural materials properties should be expected in the development of CM.

The efforts of specialists related to the development of new CM are aimed at increasing the specific strength, a significant increase in heat resistance, reducing the weight of structures, and improving their other performance.

Currently, there is no doubt that the creation of high-speed passenger trains and cars, new generation aircraft and ships without the development of new CM is impossible. The creation of new highly efficient vehicles, as a rule, is accompanied by the simultaneous development of new structural materials, primarily composite ones. As before, the most effective materials will probably appear in high-tech industries (aircraft and rocket engineering).

An important factor contributing to the composites application in the consumer goods manufacture is the cost reduction of their production. This is possible only by developing relatively simple and cheap manufactural processes. Therefore, in the coming years, we should expect great changes in this area.

Questions

(1) What word did the concept "composite material" come from?

(2) What are the main features of composite materials?

(3) What are the types of materials with structural nonuniformity?

(4) How are composite materials classified according to the common composite materials classification?

(5) How are composite materials classified according to the components geometry?

(6) How are composite materials classified according to the components space arrangement?

(7) How are composite materials classified according to the nature of components?

(8) How are composite materials classified according to the matrix material?

(9) How are composite materials classified according to the components structure and arrangement?

(10) How are composite materials classified according to the reinforcing scheme?

(11) How are composite materials classified according to the production technology?

(12) How are composite materials classified according to the application?

(13) What is the area of modern composite materials application in the aviation industry?

(14) What is the trend in the composite materials application?

Chapter 2　Composite Materials' Components

2.1　Matrix materials

2.1.1　Metal matrix

For the metal CM manufacture, the aluminum, titanium, magnesium, copper, nickel, cobalt and other materials are used as a matrix.

1. Aluminum alloys

Aluminum, as well as wrought and cast alloys based on it, are most often applied in the composites manufacture. Aluminum is applied in the CM production both by liquid and solid-phase methods. In the composites manufacture by solid-phase methods, wrought aluminum alloys are widely used. Examples of such alloys are A95052, A95154, AA5056, A92017, AA2024, 2618, 7075 (numerical tagging USA) and others. The advantages of wrought aluminum alloys are low density, high level of mechanical performance, high plasticity, etc.

To obtain composites by liquid-phase methods (casting, impregnation), cast aluminum alloys are used. These alloys are specified by increased fluidity, low linear shrinkage, and a reduced crack creation at high temperatures. Examples of such alloys are A02420, A04130, A14130, A3560, A25140, 2024, 2618, 7075 (numerical tagging USA) and so on.

The chemical composition of some matrix materials based on aluminum is given in Table 2.1.

Table 2.1　Chemical composition of some aluminum matrix materials

Material grade			Chemical composition/%				
Russia	USA	China	Al	Mg	Si	Mn	Cu
AMg2	A95052	—	95.7—98.2	1.7—2.4	0.4	0.1—0.5	0.15
AMg3	A95154	—	93.8—96	3.2—3.8	0.5—0.8	0.3—0.6	0.1
AMg5	AA5056	—	91.9—94.68	4.8—5.8	0.5	0.3—0.8	0.1
D1	A92017	—	91.7—95.5	0.4—0.8	0.2—0.8	0.4—1	3.5—4.5
D16	AA2024	—	90.9—94.7	1.2—1.8	0.5	0.3—0.9	3.8—4.9

(Continued)

Material grade			Chemical composition/%				
Russia	USA	China	Al	Mg	Si	Mn	Cu
AK4	2618	—	91.2—94.6	1.4—1.8	0.5—1.2	0.2	1.9—2.5
VD95	A97075	—	86.3—91.5	1.8—2.8	0.5	0.2—0.6	1.4—2.0
AL1	A02420	—	93.25—93.50	1.25—1.75	0	0	3.75—4.5
AK12	A04130	ZL102	87—90	0	10.0—13.0	0	0
AL3	—	ZL104	91—93.05	0.35—0.6	4.5—5.5	0.6—0.9	1.5—3.0
AL9	A03560	ZL101	91.6—93.8	0.2—0.4	6—8	0	0
AMg5K	A25140	ZL303	92.8—94.6	4.5—5.5	0.8—1.3	0.1—0.4	0
AM5	—	ZL201	93.35—94.75	0	0	0.6—1.0	4.5—5.3

CM based on aluminum also can be obtained by powder metallurgy and gas-thermal spraying. Aluminum alloys powders are used to implement these methods, for example, powder from primary aluminum, powder from secondary aluminum, dispersed powder, dispersed powder with a titanium additive.

The diffusion welding is also solid-phase methods for composites production. This method is based on the aluminum alloys foil application as a matrix material. Aluminum foil is available in rolls. The thickness of the foil is 5—200 microns. Aluminum alloys, such as AA1135 (AD1), 1199 (A99), can be used for its manufacture. High-strength complex-alloyed wrought aluminum alloys are also used for the composites production.

The explosion welding can also be applied. This technology requires sheets of aluminum [AA1135(AD1), 1199(A99), etc.] and its alloys [A93003(AMts), A95052(AMg2), AA3004(D12), etc.] with a thickness up to 2.5 mm are used. It is possible to use different sheets conditions, such as annealed conditions, semi-cold-worked condition, cold-worked condition, hardened condition and natural aging condition, etc.

2. Magnesium alloys

Magnesium is a material whose application is effective as a matrix of metal composites. It has high specific strength, low density, etc. It is very important that magnesium practically does not react with most materials applied in the reinforcing fibers manufacture. Such magnesium alloys as MAG4(ML12), MA21, AZ31B (MA2), etc. are used.

3. Titanium alloys

Titanium alloys application as a metal matrix is promising for the CM production. The high specific strength of such alloys makes it possible to use them widely in aviation and rocket engineering. Titanium alloys such as IMI125 (VT1 – 0), IMI115 (VT1 – 00), IMI318ELI (VT6S), VT22, are used for the composites production. It is used in powder or foil condition.

4. Copper alloys

Copper alloys have low level of strength properties at high temperatures. It is possible to adjust CM on the base of copper alloys by reinforcing them with fibers of tungsten, iron, and graphite.

5. Nickel alloys

Nickel-based alloys are used as a matrix material in the production of composites. The disadvantage of composite materials based on technically pure nickel is their low heat resistance. But, the scale-resistant and heat-resistant nickel alloys with the Ni-Cr system have no such disadvantage. The titanium (Ni_3Ti) and aluminum (Ni_3Al) additives application are provided with the high level of long-term strength, creep resistance, fatigue performance. Composite based on nickel alloys is created by using liquid-phase and solid-phase methods, as well as powder metallurgy methods.

6. Cobalt alloys

Cobalt-based alloys can be used as a matrix material for the composites production. Cobalt alloys are used in the form of powders, tape, melt or gas-phase methods particles. Cobalt alloys have lower strength at high temperatures compare with nickel-based alloys. For the CM production, wrought alloys are applied in the solid-phase methods and cast alloys are applied in the liquid-phase methods.

2.1.2 Polymer matrices

A CM with a polymer matrix application depends on the parts operational conditions. Many CM performances depend on the matrix, such as strength, heat and moisture resistance, resistance to an aggressive environment. A polymer CM production method depends on the polymer matrix type.

In the polymer CM production, a matrix can base on pure polymers (powders, granules, sheets, films) or as binders. The binder is a two- or multi-component system of synthetic resin and hardeners or initiators, catalysts, accelerators. Solvents, colorants, plasticizers, stabilizers and other components may be added to the binder. These components purpose is to give the binder and polymer CM the necessary manufactural and operational performance.

The following binders are most often applied in the reinforced plastics manufacture: polyesters, phenols, epoxy compounds, silicones, alkyds, polyamides, fluorocarbon compounds, acrylics, polypropylene, polyethylene, polystyrene. The thermoreactive binders are the most widely used. Heating of thermoreactive binders provides irreversible structural and chemical transformations.

1. Polyethylene

Polyethylene is one of the most widely used polymers. The industry produces high-density polyethylene (PEHD) and low-density polyethylene (PELD). The material destruction is at temperatures above 290 ℃. Solar radiation provides the polyethylene thermal aging. The strength, heat resistance, and chemical resistance of PEHD are higher than PELD. The gas permeability, quite the opposite, is higher for PELD. The polyethylene is stable to diluted sulfuric and nitric acids, concentrated hydrochloric, phosphoric and hydrofluoric acids at room temperature. For PELD, water absorption is 0.022% per 30 days; for PEHD, it is 0.005%—0.04%.

PELD is processed by injection molding ($T = 150$—200 ℃, $p = 100$ MPa), extrusion ($T = 110$—180 ℃, $p = 8$—10 MPa), pressing ($T = 130$—150 ℃, $p = 4$—10 MPa); it is well welded, machining. PEHD is processed by extrusion ($T = 180$—260 ℃), injection molding ($T = 200$—270 ℃, $p = 120$ MPa), pressing ($T = 145$—180 ℃, $p = 6$—10 MPa).

2. Polypropylene

Polypropylene supports bending well and has high worn resistance. The thermal destruction starts at 300 ℃ if the air is absent. Polypropylene is resistant to many acids and alkalis. Water absorption is 0.5% per 6 months at 20 ℃. Concentrated sulfuric acid destroys polypropylene slowly at room temperature and rapidly at 60 ℃. Polypropylene is sensitive to the oxidizer action. Polypropylene is processed by injection molding ($T = 200$—220 ℃, $p = 35$—42 MPa), extrusion, vacuum molding, blow molding, blowing, welding, pressing, spraying, and is machining by cutting.

3. Polystyrene

Polystyrene is a brittle polymer with high radiation resistance, easily aging. There are block polystyrene, suspension and emulsion. The tensile strength is 35—45 MPa. Thermal destruction starts at temperatures above 266 ℃. Polystyrene is resistant to certain mineral and organic acids, alkalis, transformer oil, and is destroyed by concentrated nitric and acetic acids. Polystyrene processes include injection molding ($T = 160$—230 ℃, $p = 80$—120 MPa), extrusion, vacuum molding, blow molding.

4. Polytetrafluoroethylene

Polytetrafluoroethylene is a linear polymer having the chemical composition $[-CF_2-CF_2-]_n$. There are polytetrafluoroethylene grades in different countries: Fluoroplast – 4 (Russia), Teflon (USA), Polyflon (Japan), etc.

Polytetrafluoroethylene has the properties of self-lubrication, and a low coefficient of friction. It is the most chemically resistant polymer, not sensitive to the fungi action. It has high resistance to the action of highly concentrated and dilute acids and alkalis, oxidizers.

Polytetrafluoroethylene is processed by sintering pre-compressed tablets. The polymer processes include machining (at high speed and light feed), welding and gluing after preliminary special processing. Some variants of polytetrafluoroethylene are processed by powder metallurgy approaches, pressing, extrusion, injection molding, spraying.

5. Polymethylenoxide

Polymethylenoxide is a linear polymer which has the chemical composition $[-CH_2-O-]_n$. Polymethyleneoxide is also called acetal or polyacetal resins. It resists fatigue and dynamic alternating loads well, and has low creep, high wear resistance. The polymethylenoxide friction coefficient for steel is 0.2—0.35. The polymer processes include injection molding and extrusion. The processing temperature is 180—240 ℃. The material is well processed by machining.

6. Polyphenylene sulfide

Polyphenylene sulfide has high heat resistance, oxidation resistance, radiation resistance. The polyphenylene sulfide chemical resistance is second only to polytetrafluoroethylene. Polyphenylene sulfide parts can be applied for a long time at 260 ℃. Its full decomposition occurs at 720 ℃. The polymer mechanical and physical properties are stable within 200 hours at 260 ℃.

Polyphenylene sulfide is processed by injection molding ($T=300—360$ ℃, $p=75—150$ MPa), pressing ($T=340—400$ ℃, $p=10—70$ MPa), plasma spraying, and fabric impregnation.

7. Polyoxyphenylene

Polyoxyphenylene is a simple aromatic polyester with a linear structure. It is known as arylox (Russia) or noril (USA). Polyoxyphenylene can be repeatedly processed without performance changes by injection molding. It has high manufacturability. The polymer is non-toxic, resistant to aggressive environments, fungal mold. It is processed by injection molding and extrusion. Thin-walled products of complex shape can be obtained from polyoxyphenylene.

8. Polyethylene terephthalate

Polyethylene terephthalate is a complex linear aromatic polyester of terephthalic acid. It is well known as lavsan. The polymer has a low friction coefficient and hygroscopicity. Parts made from it have shape stability. Polyethylene terephthalate is resistant to weak acids, mineral salts, esters, fats. The polymer is processed by injection molding.

9. Polycarbonates

Polycarbonates are polyesters of carbonic acid and dihydroxy compounds. Polycarbonates are produced with the name "diphlon". The polymer is optically transparent, resistant to ultraviolet radiation, has low hygroscopicity, action of microorganisms resistance. Parts made from it, have high dimensional stability. Polycarbonates are processed by all methods used for processing thermoplastic polymers. The processing temperature is 240—300 ℃. Parts made from polycarbonates are welded, glued, machined, joined with rivets, etc.

10. Polyarylates

Polyarylates are aromatic polyesters of diatomic phenols. They have high strength and heat resistance. In some cases, they successfully compete with structural metal materials. The thermal decomposition temperature of polyarylates is 420 ℃. These polymers are resistant to ultraviolet and ionizing radiation. Also, they are resistant to dilute mineral and organic acids, gasoline, kerosene at prolonged exposure. Polyarylates are processed by injection molding, pressing, extrusion, blow molding.

11. Polyoxybenzoates

Polyoxybenzoates have high heat resistance and do not melt up till reaching the decomposition temperature (about 550 ℃). The crystal structure of the polymers is kept up till the temperature of 530 ℃. Polyoxybenzoates are wear-resistant materials that are resistant to corrosion. They are effective to the polymer CM manufacture that operate at the temperature about 300 ℃. Parts made of polyoxybenzoates are produced by sintering a powdered polymer.

12. Polyimides

Polyimides are produced in the form of the granular molding compound. The polyimides performance is stable over a wide temperature range (from -200 ℃ to $+300$ ℃). The polymer friction coefficient for steel is 0.05—0.17. Products made from polyimides have high dimensional stability, low creep at high temperatures, high heat resistance and gamma radiation (also, fast electrons and neutrons radiation) resistance. Diluted acids have almost no effect on polyimides. Polyimide granular molding compounds are processed by direct compression, compression pressing, injection molding, hot pressing.

13. Polyamides

Polyamides are well known as caprolon, nylon, etc. These polymers have high fatigue strength, wear resistance, impact strength, low hygroscopicity, stability of properties at high temperatures. Polyamides are resistant to organic solvents. Aromatic polyamides are processed by direct compression with preheating. The main producing methods are injection molding and extrusion.

14. Epoxy resins

Epoxy resins are oligomers or polymers containing at least two epoxy or glycidyl groups in the molecule. They can turn into polymers of a spatial structure. Uncured resins are thermoplastic viscous fluids or brittle solids. Resin hardeners are monomeric, oligomeric and polymeric compounds of various classes. Epoxy resins cure both without heat and when heated, even in water. Resins are highly resistant to alkalis, salts, oxidizing agents, organic solvents. The industry produces epoxy-diane resins, epoxy-novolac resins, cycloaliphatic epoxy resins. Polymeric CM made on the base of epoxy resins have high mechanical properties.

The parts manufacturing based on epoxy resins are impregnating fibers, fabrics, nonwoven fibrous materials, paper, etc. After curing, they are processed by direct compression, contact molding, vacuum molding, etc. Processing temperature is 20—180 ℃.

15. Phenol-formaldehyde resins

Phenol-formaldehyde resins are polymers that based on polycondensation of phenols with formaldehyde. There are resol (thermoset) or novolac (thermoplastic) phenol-formaldehyde resins with depending on the polycondensation conditions.

In the uncured state, phenol-formaldehyde resins are a brittle transparent amorphous mass that transforms into a liquid state at 60—120 ℃. The resole resins performance changes over time, novolac resins performance is stable in the moisture absence.

The curing resins maximum temperature is in the range of 140—200 ℃. Resin curing products are resites. These are fragile materials with a non-crystalline structure, high strength, electrical insulating, anti-corrosion properties. Resites are resistant to most acids. The destruction temperature of phenol-formaldehyde resins is higher than 300 ℃.

The phenol-formaldehyde resins processing based on the impregnation of woven and non-woven fibrous fillers, paper, etc. , and after curing, they are processed by direct pressing, injection molding, extrusion and vacuum molding.

16. Furan resins

Furan resins are oligomeric compound containing a furan ring in molecules and capable of being converted into three-dimensional polymers at these conditions: catalyst presence and (or) upon heating. Furan resins cure in the same way as phenol formaldehyde. Furan resins based materials are heat-, acid- and alkali-resistant, their decomposition starts at temperatures above 300 ℃. The processing of furan resins into polymer CM consists of impregnating fillers such as fabrics, fibers, etc. They are processed by direct compression, autoclave molding, etc. after curing.

17. Organosilicon polymers

Organosilicon polymers (silicones, poly-organosilicones) are heat-resistant high-molecular organoelement compounds. Their macromolecule contains silicon and carbon atoms. Polymers are resistant to most acids and alkalis. Resins are processed by impregnation of fillers followed by curing. The composites are manufactured by direct pressing, contact molding, etc.

2.1.3   Ceramic matrices

Currently, the term "ceramics" has several meanings. So, there are ordinary and technical ceramics. Ordinary ceramics contain silicates (SiO_2), therefore, the industry that produces ordinary ceramics is sometimes called silicate. The special-purpose ceramics (technical ceramics) are applied for the engineering, which includes various oxides, carbides, nitrides, borides, silicides, sulfides. Also, mixed-type ceramics present. For example, on the base of two types of ceramics: ionic oxide Al_2O_3 and covalent anoxic Si_3N —an effective material "sialon" with the general compound $Si_{6-x}Al_xN_{8-x}O_x$ was created. This type of ceramics can be applied in internal combustion engine cylinder blocks, gas turbine blades, etc.

In some cases, cermets materials are qualifying classified as ceramic materials. Cermets (ceramic-metal materials) are produced by powder metallurgy methods. They include ceramic (more than 50% by volume) and metal components. The alumina oxide cermets —metals are most widely applied cermets, for example chromium used as metal. Besides the alumina oxide, oxides of magnesium, beryllium, titanium, zirconium, chromium and other elements can be used as a ceramic component in cermets. The metal component is usually nickel, cobalt, iron, and some metal alloys.

　　Sometimes ceramics is understood as a solid substance that has the nonmetallic nature of the linkage and non-polymer (chain) structure, i. e. , ceramics is everything that is not a metal or a polymer.

　　There are ceramics by the functional area: mechano-, thermos-, electro-, magneto-, opto-, chemo-, bio-, nuclear- and superconducting ceramics. In this book, mechano-ceramics and thermos-ceramics will be described.

　　The main performance of mechano-ceramics are hardness, strength, elastic modulus, fracture toughness, wear resistance, tribotechnical properties, linear thermal expansion coefficient, heat resistance. These properties are provided by the compounds: Si_3N_4, ZrO_2, SiC, TiB_2, ZrB_2, TiC, TiN, WC, B_4C, Al_2O_3, BN.

　　Thermal ceramics have thermal integrity, heat resistance and fire resistance. As the ceramic components, SiC, TiC, B_4C, TiB_2, ZrB_2, Si_3N_4, BeS, CeS, BeO, MgO, ZrO_2, Al_2O_3, TiO are used. The performance of some types of oxide and oxygen-free ceramics are given in Table 2. 2.

Table 2. 2　Performance of some types of oxide and oxygen-free ceramics

Ceramics	Density,ρ/ $(kg \cdot m^{-3})$	Melting point/ $°C$	Mohs' hardness	Ultimate stresses/MPa	Elastic modulus E/GPa
α-Al_2O_3 (corundum)	3,900	2,050	9	140—265	350—490
ZrO_2 (stabilized ceramics)	5,600	2,500—2,600	7—8	148—130 (20—1,500)	172—96 (20—1,350)
BeO	3,000	2,570	9	105—280 (20—1,200)	274—70 (20—1,200)
MgO	3,580	2,680	6	98—42 (25—1,300)	300—220 (25—1,300)
SiC (carborundum)	3,200	2,600	9. 2—9. 5	155	394
ZrB_2+10 % Mo_2Si	5,500	2,370	—	—	450
BN	2,340	2,350	1—2	50—110 (25)	—10 (1,000)
Si_3N_4 (hot-pressed silicon nitride)	3,190	1,900	—	—	317
$MoSi_2$	6,240	2,030	—	60 (1,100)	430

　　The temperature range ($°C$) is given within brackets, at which the performance were measured.

There are examples of ceramic materials that have been successfully used: silicon carbide (for engines nozzle inserts); titanium carbide (thermal integrity materials for the parts of jet craft); zirconium diboride (aircraft surface and engine nozzles edges); boron nitride (fairings of aircraft electronic equipment).

Technical ceramics is the third most frequently applied industrial material. The estimated annual economic effect of the technical ceramics application in the United States is $ 3 billion.

The main advantages of ceramic materials are their high melting points, high strength properties under compressive stress, chemical resistance.

The main disadvantage of technical ceramics is the crack resistance low level (fracture toughness). The fracture toughness of ceramics is about 1—2 MPa · $m^{-0.5}$. For metallic structural materials, the fracture toughness is about 40 MPa · $m^{-0.5}$. The low crack resistance of ceramics depends on the chemical linkage high strength. Covalent linkage materials are typically strong and brittle.

Methods for the technical ceramics crack resistance increasing are based on the optimization of the initial powders structure, the material molding and sintering manufactural approaches. These are methods which are traditional for materials science.

The crack braking methods under load provides a crack resistance increasing. This approach is applied in ceramic materials based on zirconium dioxide ZrO_2. The initial structure with tetragonal ZrO_2 compound is converted to monoclinic ZrO_2 compound under pressure application. Monoclinic modification of ZrO_2 has a volume of 3%—5% greater than tetragonal. Therefore, if a crack grows in these ceramics, then the additional volumes of monoclinic modification ZrO_2 are expanded and the crack will be compressed. The ceramics crack resistance grows till about 15 MPa · $m^{-0.5}$. Schematically, this effect is shown in Figure 2. 1 (a).

Another method for crack resistance increasing of technical ceramics is based on the voids (discontinuities) creation in the ceramic matrix. The void radius is much larger than the radius at the crack tip. If the crack meets the void, then its top becomes obtuse [Figure 2. 1 (b)]. As a result, the material crack resistance increases. By the voids geometric parameters controlling (their shape and volume fraction) it is possible to effectively increase the ceramics crack resistance level.

An additional method of crack resistance increasing of technical ceramics is to reinforce the material with special fibers. High-strength fibers are located perpendicular to the crack propagation direction and support tensile stresses. The fibers actually provide a crack resistance increasing of the ceramic matrix by preventing crack propagation [Figure 2. 1 (c)].

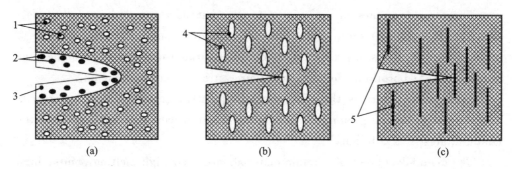

1 —tetragonal ZrO_2; 2 —monoclinic ZrO_2; 3 —compresses stresses zone; 4 —voids; 5 —fibers

Figure 2. 1   Crack resistance increasing approaches of the technical ceramics

Despite the high brittle of most ceramic materials, some of their grades are used as impact resistant. It is possible because ceramics have high values of hardness, elastic modulus, melting point, low density.

Ceramic and cermet matrices provide the ability of composites application at high temperatures zones. Matrices of SiC and Si_3N_4 can be applied at operational temperature up to 1,700 ℃. Oxide ceramics with super-refractories is produced by processing of sintering fine-dispersed powders of high-temperature oxides: aluminum, zirconium, beryllium, magnesium, calcium, etc. The melting points of the oxide ceramics crystalline phase are in the range of 2,000—3,300 ℃. The operational temperature is about (0.8—0.9) of melting point. Polycrystalline oxide ceramic has high thermal conductivity and refractoriness, thermal and chemical resistance, elastic modulus. Super-refractories oxides are stable in dry air and in oxidizing environments.

Chemical and thermal resistances are the most important advantages of ceramic materials. These qualities are provided by strong chemical interatomic linkage. If we talk about oxide ceramics, then it can be considered as a material that has already "burnt" and "corroded". Therefore, further oxidation of such ceramics, for example, during combustion or other chemical reactions, is practically impossible.

The oxidation resistance of oxygen-free ceramics is insufficient at temperatures above 900—1,000 ℃. An exception is oxygen-free compounds containing silicon, since a protective film of SiO_2 is formed on the material surface, when it operates in an oxidizing environment.

The main stages of ceramic production are the powders production and their consolidation. The powder particles consolidation is provided by the processing of material high-temperature heating. The powder production is the most important stage of ceramic manufacturing. High-quality ceramic is produced by using fine-dispersed initial powders (with a particle size of up to 1 micron).

Technical ceramics powder molding is provided by various approaches: pressing, extrusion (extruding the molding material with plasticizer additives through the orifice), slip casting. Parts with simple shape are produced mainly by hot pressing with homogenizing annealing. The hot pressing temperature is 80%— 90% of the compound melting point. Parts with complex shape are produced by extrusion or slip casting. A future-oriented technology for molding ceramics is explosive and hydrodynamic pressing.

The sintering ceramics process provides the material porosity decreasing and part strength sharp increasing. Sintering is the reason for material parts capping. The ceramics sintering temperature is about 2,000 ℃. In some cases, the molding process combines with the ceramics sintering process.

X-ray and ultrasonic inspection are usually applied in ceramic materials quality control. Ceramics processing and quality control determine the parts' price. In some cases, the initial materials cost and their consolidation is only 11%, the ceramics part processing is 38%, and their quality control is 51% of the total cost.

2.2   Reinforcing elements

Reinforcing elements are the most important components of CM. Many materials in various structural conditions are applied to reinforce composites. The chemical composition, structural state and geometric parameters of the reinforcing elements depend on the material requirements. The matrix type selected for CM is very important.

For the polymer composite materials production, glass, carbon, boron and organic fibers are most often applied. A fibers comparative performance is listed in Table 2. 3. The following reinforcements are used for metal matrix composites: thin steel wires, tungsten wires, beryllium, niobium, titanium, etc. Tensile diagrams of some types of high-strength reinforcing fibers are shown in Figure 2. 2.

Table 2. 3   Fibers comparative performance applied in CM

Performance	Fiber types			
	Glass	Boron	Carbon	Aramide
Mechanical performance:				
Specific strength	High	High	Medium	Extra high
Specific modulus	Low	High	Extra high	Medium
Shock resistance	High	Medium	Low	High
Tensile strain	High	Low	Medium	Medium

(continued)

Performance	Fiber types			
	Glass	Boron	Carbon	Aramide
Stability	High	High	Medium	High
Thermal and physical performance				
Thermal conductivity	Low	Medium	High	Low
Thermal coefficient of linear expansion	Medium	Medium	Very low	Very low
Dampening quality	High	Low	Medium	High
Manufactural performance				
Minimal bend radius	Small	Extra large	Small	Small
Processing vulnerability	Medium	Medium	High	Low
Tape and fabric processing	Good	Bad	Good	Bad
Price	Very low	High	Medium	Medium

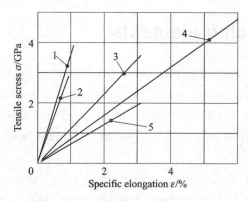

1—boron fibers; 2 —high modulus carbon fibers; 3 —organic fibers;
4—high stress glass fibers; 5 —electroinsulating glass fibers

Figure 2. 2   Typical tensile diagrams of high-strength fibers applied in CM

2.2.1   Metallic fibers

One of the cheapest types of reinforcing elements for CM is metal wire. The metal wire production is a well-known process. Get it by drawing. The metal wire properties of different grades provide its efficient application. The mechanical performance of some types of metal fibers are given in Table 2. 4. The metal wire has higher manufacturability compared with other fibers types. Small diameter wire can be processed to gauze (by applying textile approach) and applied for CM reinforcing. The wire from high-carbon and low alloy steels, stainless, martensitic-aging steels, tungsten, molybdenum, titanium, niobium, etc. is often applied for

CM reinforcing.

Table 2.4　Metallic fibers performance

Material	Density $\rho \times 10^3 / (\mathrm{kg} \cdot \mathrm{m}^{-3})$	Diameter $d/\mu \mathrm{m}$	Average strength σ_1^+ /GPa	Elasticity modulus E_1^+ /GPa
Aluminum	2.7	—	0.29	70
Beryllium	1.85	130	1.1	310
Titanium	4.5	—	0.55	120
Silicium	2.5	—	1.0	72
High strength steel	7.8	100—300	3.5—3.8	200
Molybdenum Vanadium (+5 %)	—	250	1.8—2.0	334
Tungsten	19.3	50	3.3	410

A steel wire has a high level of mechanical performance. A wire with a diameter of about 20—1,500 microns is most often used. The cheapest wire is made from carbon steel. The high-strength stainless steel wire has high efficiency at the operational temperature in the range of 77—623 °K. Stainless steel wires interact weakly with matrix materials, because they have the relatively high heat resistance and a passive surface.

The martensitic-aging steel wire application has high efficiency for CM based on light metals and alloys. Martensitic-aging steels hardening process is at temperatures of about 480—500 ℃.

The high-strength carbon steel wire manufacturing process consists of patenting and subsequent cold drawing.

The strength of the small diameters wire depends on its value. The smaller the diameter is, the higher the strength properties of the wire are. For example, the wire tensile strength made of steel grade W1-7 (USA) with a diameter of 80 microns is about 4,400 MPa, but with a diameter of 260 microns is about 3,600 MPa. With a wire diameter increasing, the tensile strength sensitivity to this factor decreases. High-strength steel wire is sensitive to defects in the internal structure and external surface. Therefore, with wire length increasing, its strength average level decreases.

For the CM production, metal wire is applied in the form of separate fibers or gauze. Woven gauze can be made only from high plastic materials, which, as a rule, have low strength. Their other disadvantage is the pinches creation at the contact points of the warp wire and the weft wire.

　　The tied-wire fabric doesn't have these disadvantages. There are gauze of the type "stockinette structure", "rib", "full-cardigan structure", "half-cardigan structure". High-strength wires with a specific strain of 2%—3% with a diameter of 20—200 microns are suitable for the tied-wire fabric. The tied-wire fabric volumetric structure provides a low level of contact stresses, as a result, the tied-wire fabric destruction does not occur at the wires contact points, but along their length.

　　The composites based on copper, nickel, cobalt, titanium and their alloys, tungsten and molybdenum wires are used as reinforcing fibers. They have high strength and rigidity. The tungsten elastic modulus is second only to osmium, iridium, and rhenium. High strength properties of tungsten and molybdenum are both for regular and for high temperatures.

　　The manufactural processes for these wire types production is based on the tungsten and molybdenum powders application. The tungsten wire minimum diameter is about 20 microns. Tungsten and molybdenum wire with a diameter of 20—80 microns are applied in the gauze and tied-wire fabric manufacture.

　　It is more important to provide the components thermodynamic compatibility, i. e., to prevent the diffusion and chemical interaction between the fiber and the matrix for the metal CM production. The set number with thermodynamic component compatibility is limited. The most active interaction between the fiber and the matrix is during the liquid-phase methods application. In addition, the components active interaction takes place during the CM parts using at high temperatures.

　　Metal composites based on the nickel matrix reinforced with tungsten and molybdenum fibers can be used at 1,000—1,200 ℃. But, at such temperatures, the components thermodynamic stability is absent. The undesirable interaction is between the CM components. The antiphase boundary stability (between the fiber and the matrix) can be increased by applying fibers anti-diffusion coatings. Titanium nitride, thorium, aluminum and zirconium oxides are recommended as protective coatings for tungsten and molybdenum fibers.

　　The beryllium wire is applied like discrete and continuous fibers for CM reinforcing. Its main features are high elasticity modulus and high specific strength. The tensile strength of the wire with a diameter of 51 microns is 1,455 MPa. The beryllium wire high rigidity does not allow the gauze production.

　　Titanium wire has high specific strength and high corrosion resistance. The tensile strength of the wire with a diameter of 800 microns made of VT9 alloy (Russia) is 1,820 MPa. Titanium wire is applied in the form of continuous and discrete fibers for CM.

Bimetallic wire can also be used for CM. The cladding layer volume fraction in it can be 20%—40%. The manufactural process is next: bimetal blank creation (by casting, welding, soldering, metallizing, electroplating, etc.) and its subsequent drawing. The bimetallic wire mechanical performances are depended on the used materials types and their balance. For the composites manufacture, such wires are applied like fibers, gauze and compound in the CM production by fiber placement with followed compaction by explosion, welding, pressing.

There are other manufactural processes for the metal fibers creation. One of them is based on the melt creation and extrusion of it through small holes. Liquid metal is extruded through multihole spinnerets. Then, the fibers are cooled in the chamber by air or inert gas.

The Taylor method is used for small diameter metal fibers production (less than 50 microns). It is based on the combined drawing of metal and glass. This method allows to obtain fibers with a diameter of 1—30 microns. Glass insulation is erased by stripping method in acids and molten alkalis from metal fibers.

2.2.2   Glass and silica fibers

Fiberglass is one of the most used CM. In 1979, more than 900 thousand tons of this type of materials were produced in the USA. Such a large volume of fiberglass production is due to their high performance and relatively low cost.

For the fiberglass production, glass fibers of various types are required. Now, a large number of fibers grades and their production effective manufactural processes have been developed.

The basis of glasses is silica (SiO_2). The melting point of silicon dioxide is very high. Various additives can be applied into the glass to reduce it. The additives can change the part's performance. Glass is an amorphous material that does not have a crystalline structure. According to its physical and mechanical performance, it has a position between solid and liquid.

Glass fibers' widespread applications are based on the wide range of advantages. The fiber density is relatively low and is in the range of 2.4×10^3—2.6×10^3 kg · m^{-3}. Glass fibers have a high level of tensile stresses and very good electrical insulators. Their specific strength is higher than that of steel wire. Glass fibers, being inorganic materials, do not fire. The fibers' high melting point allows them to be used at high temperatures. Glass fibers have a low linear expansion temperature coefficient and a high thermal conductivity coefficient. They are chemically resistant, are stable to most chemicals, are resistant to fungi, bacteria and insects. Glass fibers have increased moisture resistance, do not swell on wetting or moisture, and keep their own performance in high humidity environments.

Typically, the glass fibers cross-sectional shape is a circle. However, for specific cases, hollow and shaped fibers are produced with the cross-sectional shape: triangle, square, hexagon and rectangle (Figure 2.3). Glass fibers are used like continuous fibers and cut staple.

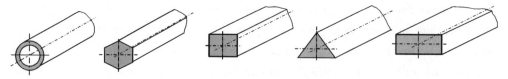

Figure 2.3　Glass fibers shapes

The glass fibers' manufactural process consists of the following stages: preparing silica sand, limestone, boric acid, clay, coal, fluorite and other components, mixing and melting them in high-temperature furnaces. The composition melting point is approximately 1,260 ℃. There are one- and two-step processes for fiberglass production. At the one-step process, the glass melt enters directly into the equipment, which makes glass fibers, yarn, or cut staple. At the two-step process, the melt is first processed into glass spheres, which then transfer to the melting furnaces and are further processed.

The glass fibers' diameter is about 3—19 microns. The diameter depends on the wind-up speed or the jet velocity. For the glass fibers production, many processes parameters are monitored; the most important of which are the material heating temperature, melt viscosity, etc.

The glasses performances depend on their chemical composition. Typical (commercial) glass grades are A, C, E and S. A-glass, highly alkaline (sodium, bottle), is the most common. A high alkali content determines their low electrical performance. E-glass has good electrical insulating performance, because they are low-alkaline. E-glass is main for the fiberglass textile.

C-glass has high chemical resistance. S-glass is high strength. The S-glass fibers' strength is about 40% higher than the E-glass fibers' strength. The glasses chemical composition of grades A, C, E and S are given in Table 2.5.

Special glass grades are M-, D - and L-glasses. M-glass is highly modular. Their Young's modulus is about 113 GPa. They have high carcinogenicity, because the beryllium oxide is in their composition. D -glass has low dielectric properties. Their application can be effective for the antenna and radars fairings. L-glass is lead-containing and applied in radiation protection.

Table 2.5 Glasses chemical composition

Composition	Glass grades			
	A-glass (highly alkaline)	C-glass (high chemical resistance)	E-glass (electrical insulating)	S-glass (high-strength)
Silicon oxide	72.0	64.6	54.3	64.2
Alumina oxide	0.6	4.1	15.2	24.8
Iron oxide	—	—	—	0.21
Calcium oxide	10	13.2	17.2	0.01
Magnesium oxide	2.5	3.3	4.7	10.27
Sodium oxide	14.2	7.7	0.6	0.27
Potassium oxide	—	1.7	—	—
Boron oxide	—	4.7	8.0	0.01
Barium oxide	—	0.9	—	0.2
Other components	0.7	—	—	—

Glass fibers' performances are determined by their chemical composition and structure. The performances of magnesium aluminosilicate, aluminumborosilicate and acidresistant glass fibers are listed in Table 2.6.

Table 2.6 Glass fibers' performances

Performance	Glass grades		
	Magnesium aluminosilicate	Aluminumborosilicate	Acidresistant
Physical:			
Density/(kg · m^{-3})	2,480	2,540	2,490
Mechanical:			
Ultimate tensile stresses σ_1^+/MPa, at:			
22 ℃	4,585	3,448	3,033
371 ℃	3,768	2,620	—
533 ℃	2,413	1,724	—
Tensile elasticity modulus E_1^+/MPa, at 22 ℃	85.5	72.4	69.0
Yield point/MPa	5.7	4.8	4.8

(continued)

Performance	Glass grades		
	Magnesium aluminosilicate	Aluminumborosilicate	Acidresistant
Thermal:			
Coefficient of Linear Thermal Expansion $\alpha \times 10^6 / \mathrm{K}^{-1}$	5. 6	5. 0	7. 2
Thermal conductivity $\lambda / \mathrm{Wt} \cdot (\mathrm{m} \cdot \mathrm{K})^{-1}$	—	10. 4	—
Specific heat $c_m / \mathrm{J} \cdot (\mathrm{kg} \cdot \mathrm{K})^{-1}$, at 22 ℃	0. 176	1. 197	0. 212
Softening point $T / ℃$	—	841	749

In addition to the noted glass fibers types that have a complex chemical composition, there are quartz and high silicate fibers. Quartz glass is produced from natural quartz crystals and contains at least 99. 95% SiO_2. There are few deposits in the world of pure natural quartz that can be used for high-quality fibers production. In particular, mines producing such quartz present in Brazil. The advantages of quartz fibers are high strength and heat resistance. The quartz fiber elasticity modulus increases with temperature increasing: from 74 GPa (at 300K) till 83 GPa (at 1,200K). High-silica and quartz glasses have high thermal shock resistance.

Quartz glass does not lose high viscosity even at the temperature of 2,400K. The main way to fibers creation from quartz is drawing from heated rods. The original quartz rods diameter is 0. 2—2. 0 mm and 4. 4 mm.

Common scheme for manufactural process for quartz rod production is shown in Figure 2. 4.

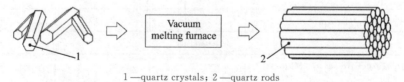

1 —quartz crystals; 2 —quartz rods

Figure 2. 4　Manufactural process scheme for quartz rod production

The main raw material for the high-silicate glass production is high-purity sand. Silica (high silicate) fiber contains 95%—99% SiO_2 and it is product from silicate glasses by process of leaching technology. The high-silicate glass production main stages:

1) silicate glass production with a content of about 65% SiO_2;

2) melting of silicate glass;

3) fiberglass production;

4) leaching in dilute mineral acids solutions;

5) washing with water;

6) drying;

7) heat treatment at 540 ℃ (to remove linked water).

Single continuous fiberglass strands, which are produced directly at the bushing exit, are the simplest form of textile yarn ("simple yarn"). The simple yarn is pre-twisted and after that is textile processing. Thin simple yarn is rarely applied in practice. More often, a thicker yarn is needed, which can be produced by twisting two or more strands. Subsequent twisting of two or more pre-twisted strands is called slubbing.

It is necessary to apply an air stream, in order to create the bulky fiberglass yarn effect. In this case, the elementary fibers that are located on the yarn surface are randomly destroyed, and yarn will be garneting. This creating bulky yarn process is called crimping. The process is controlled by air pressure and yarn feeding speed.

The roving are produced by together spinning of several simple strands. Most of them are processed into coarse cloth. The advantage of roving fabrics is the ability to rapidly set required thickness.

Part of the fiberglass is used to mats production. Fiberglass mats are produced from cut or continuous strands, as well as from simple continuous. Mats made from cut fibers can have a thickness till 2 m. The fiberglass mats that created from continuous monofilament yarn with very small thickness can be used as decorative surface-reinforced layers.

Fiberglass yarn can be applied in the fiberglass cloth production. Fiberglass cloth is used for the polymer composites production. They are usable for the bulky parts production.

There are two yarns types of cloth: warp and weft. The yarns oriented along the cloth are called the warp; the weft tangles the warp in the crosswise. There are several textile clothes types which depend on the warp and weft tangle types. Fiberglass clothes can have tabby weave, in which the weft placed sequentially under warp and above it. The advantages of this type of cloth are the low damageability of the yarns and the high stability when slipping yarn.

There are tabby, twilled, satin and other types of weaves for the fiberglass cloth production (Figure 2.5). The industry mainly produces plain and satin weave clothes. In addition, fiberglass yarn is used for the tape, outline cloth, goffered cloth, 3-d cloth, etc. manufacture.

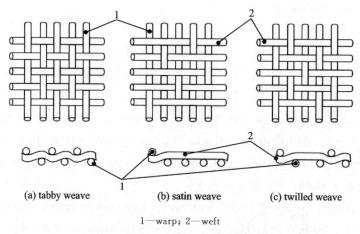

(a) tabby weave   (b) satin weave   (c) twilled weave

1—warp; 2—weft

Figure 2.5　Fibers cloth weaves types

Quartz and highly silicate fibers have higher strength and flexibility than regular glass fibers, so they have higher manufacturability too. Due to their high performance, highly silicate and quartz fibers are processed into almost all textile types.

The cloth weave type and density influence on its flexibility, molding capacity, damaging, strength and determine the bonding impregnation ability. Tight-textured weave clothes have low molding capacity and bonding slowly impregnated. It is one of the reasons for the low strength of the CM from them. Clothes with open weave also do not provide a high strength. In each case, the weave rational type is determined on specific conditions.

For strength and wear resistance increasing, the metallization finish, for example nickel, copper, iron or zinc, can be applied to the glass fibers surface. Metallization finish is applied by the vapor phase or the liquid bath metallization.

The fiber content in fiberglass CM can reach 80%. Such a fiberglass high content and, accordingly, a blinding low volume part in the composites are achieved by materials rolling in their production. It is necessary to use close-packed methods for fiber volume part increasing. The fiberglass optimal content in CM is 30%—32%. The fiberglass mechanical performance which depends on the blinding part volume is shown in Figure 2.6.

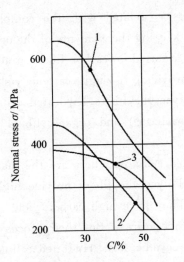

1—banding; 2—tensile; 3—compression

Figure 2.6   The fiberglass strength dependence on the blinding part volume C

2.2.3 Carbon fibers

The next reinforcing fibers type that will be explained in this book are carbon fibers. The carbon fibers application scope is constantly increasing in the composite materials industry, due to the high level of their performance. Their specific indicators are higher than all heat-resistant fibers. The high-modulus carbon fibers have tensile strength 2.5—3.5 GPa. The modulus of elasticity, E is 200—700 GPa and a density is 1,600—1,800 kg · m^{-3}. The carbon fibers density is lower than graphite (2,260 kg · m^{-3}). This difference is based on the high fibers porosity. Carbon fibers have high chemical resistance to the most aggressive environments.

Due to the mechanical performance high level, low density, high heat resistance, etc. carbon fibers are widely used for CM with polymer, carbon, ceramic and metal matrices.

There are two types of carbon fibers with depending on the processing conditions: carbonized and graphitized. The heat treatment temperature of carbonized fibers is 900—2,000 ℃, the carbon content in them is 80%—99%. The heat treatment temperature of graphitized fibers is 3,000 ℃, the carbon content in them is more than 99%. The carbonized fibers are usually used for graphitization. Carbonization and graphitization are processed in vacuum, in various environments (methane, nitrogen, argon, etc.), as well as in coal, coke and graphite filling.

The carbon fibers production is based on the polymers heating at an inert medium and their thermal destruction. During the polymers decomposition, volatile products are formed and a solid coke residue remains. The organic fibers conversion to carbon fibers is related with the complex reactions, structural fundamental changes while keeping the elements of the original polymer skeleton.

Only fibrous polymers are used for carbon fibers production. They do not melt during heat treatment, providing a high carbon yield and high mechanical performance. They are polyacrylic fibers, hydrated cellulose fibers, carbon-rich pitches (ordinary and mesophasic) and organic fibers produced from phenolic resins.

Polyacrylic and hydrated cellulose fibers are the main raw materials for the carbon fibers production. Polyacrylic fibers provide high-strength high-modulus carbon fibers production. They have high carbon yield and simpler manufacture (lower material heating temperature). But their processing relates with a toxic substance extraction —hydrocyanic acid. Hydrated cellulose fibers are free from this disadvantage, and they are also cheaper than polyacrylic fibers.

Synthetic viscose (hydrated cellulose) fibers are the main raw material for the carbon fibers production. The manufactural process consists of the next stages: textile preparation, oxidation, carbonization, graphitization.

Textile preparation is necessary to remove moisture, inorganic impurities and organic substances, including lubricants. The treatment is carried out with thinners or superficially active substance. Subsequently, pulp is dried for 15 hours at 100 ℃.

At the pulp oxidation stage, which is carried out at temperatures of 350—400 ℃, there is large material mass losses. Carbon content is no more than 60%—70% after this operation. Further, fibers carbonization is carried out at temperatures of 900—1,500 ℃. The material is further enriched in carbon.

The graphitization is the fibers heat treatment final stage. It is carried out at an argon medium and has duration only a few minutes. The initial temperature of this stage depends on the final carbonization temperature. The final temperature is about 2,600—2,800 ℃. The fibers carbon content after the graphitization exceeds 99%.

If the cellulose fibers converting process to combine with the drawing operation, then the fibers strength and elastic performance can be improved.

The carbon fibers structure is shown in Figure 2.7. There are three layers: surface layer A, a highly oriented layer B, a low-oriented layer C. The fibers specific feature is their fibrillar structure. Voids take about 33% of the fiber volume; they are oriented along the fiber axis and have a needle-like shape.

The producing carbon fiber processes stages for hydrated cellulose fibers and polyacrylic fibers are similar.

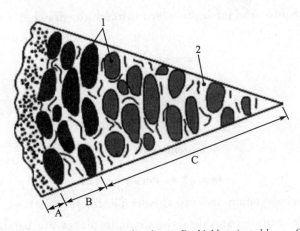

1—microfibrilles; 2—amorphous carbon; A—surface layer; B—highly oriented layer; C—low-oriented layer

Figure 2.7　Carbon fibers structure

Coal and oil pitches are complex oligomeric compound mixtures. The pitches composition and properties depend on their origin and can vary widely. Pitches are a relatively cheap raw material for the carbon fibers production. The carbon fibers production process consists of the following stages: pitch preparation; fiber spinning; long hardening; carbonization in an inert gas; fiber graphitization under load. Fibers from the pitch are produced by melt passing at 370—620 K through draw holes with a diameter of 0.3 mm. Then there is a fibers drawing of rating of 100,000%—500,000%. The carbonization, graphitization and processing stages of hydrated cellulose and polyacrylic fibers are similar. The pitches disadvantage is the high carcinogens content.

If the fiber production process includes the fiber drawing stage during graphitization and the optimal temperature and time parameters are provided, it is possible to obtain high mechanical performance fibers. The fibers with the tensile strength of 2,585 MPa and an elastic modulus of 480 GPa were experimentally created. However, all these conditions can provide a significant material cost. If the processes of fibers drawing and graphitization are not at the same time their elastic modulus cannot exceed 35—70 GPa. Such fibers are used only as a raw material for the boron deposition.

A strength low level is specific primarily for isotropic structure pitches. Liquid crystalline pitches, i.e., pitches that are mesophase state, can provide a high level of carbon fibers mechanical performance. The pitches are transferred to the liquid crystalline (mesophase) state by thermal treatment. In the mesophase state, part of the compound of particles is ordered and a liquid crystal structure forms in the material. During the subsequent the mesophase pitch processing through the draw holes, the particles are oriented along the fiber axis. The structures of the isotropic melt, mesophase pitch and pitch processed through the drawholes are schematically shown in Figure 2. 8.

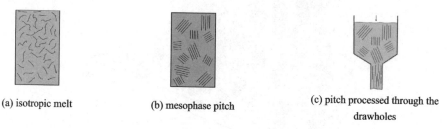

(a) isotropic melt   (b) mesophase pitch   (c) pitch processed through the
                                                              drawholes

Figure 2. 8   Pitches structure

There are radial, onion and radial-isotropic fibers structure. All these three variants have one common specific: the graphite planes are parallel to the fibers axis. The carbon fibers cross-sectional structures produced from mesophase pitches are shown schematically in Figure 2. 9.

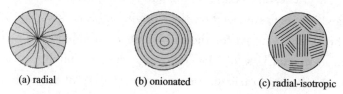

(a) radial   (b) onionated   (c) radial-isotropic

Figure 2. 9   Fibers structure produced from pitches

Medium-strength carbon fibers with a low elasticity modulus are produced from phenolic resins. The phenolic resins synthesis is simpler compared to pitches processing. Phenolic resins are useful for the carbon fibers production with a developed surface or hollow structure. The refractory and heat-resistant compounds are added into the spin dope at the carbonization and graphitization stages for the carbon fibers performance modifying. As for another way, the fibers (or cloth) are impregnated with such compounds. The result is fibers or clothes with special coatings, for example carbides or oxides.

The carbon fibers mechanical performance depend on the raw fibers structure, heat treatment specific, and final structure defects. The carbon fibers performance range can be very wide, and it depends on the internal and surface defects modes.

In general, carbon fibers have a high level of performance. By strength, carbon fibers are divided into three groups: low-strength fibers (tensile stress smaller than 500 MPa), medium-strength fibers (tensile stress equal 500—1,500 MPa) and high-strength fibers (tensile stress higher than 1. 5 GPa). The elasticity modulus is proportional to their strength and varies over a wide range.

For the CM production, carbon fibers with a high-level mechanical performance are mainly used. Conventionally, these fibers are divided into high-strength fibers and high-modulus fibers.

High-modulus fibers: elasticity modulus is 300—700 GPa; tensile stress is 2—2.5 GPa. High-strength fibers: elasticity modulus is 200—250 GPa; tensile stress is 2.5—3.2 GPa. Some fibers have high tensile strength and elastic modulus at the same time. Strength and elasticity modulus of some high-strength and high-modulus fibers are listed in Table 2.7.

Table 2.7   **High-strength and high-modulus fibers performance**

Fiber grades	Diameter/μm	Density /(kg · m^{-3})	Elasticity modulus /GPa	Tensile stress /GPa
VMN – 4 (Russia)	6.0	1,710	270	2.21
Culon (Russia)	—	1,900	400—600	2.00
LU – 2 (Russia)	—	1,700	230	2.0—2.5
Ural – 15 (Russia)	—	1,500—1,600	70—80	1.5—1.7
TORAYCA T1100G (Japan)	—	—	324	7.0

Carbon fiber distinctive materials have very high heat resistance. The carbon fibers tensile strength and elastic modulus do not change till the temperature of 1,500 ℃, when they heated in an inert medium. The temperature limit of carbonized fibers is 300 ℃ and graphitized is 400 ℃ for the long duration in the atmospheric. The fibers are protected by special coatings of pyrocarbon, oxides, nitrides and refractory carbides to prevent the surface from oxidation.

The carbon fibers thermal conductivity coefficient is approximately similar with metals. Pyrocarbon application for the carbon fiber cloth provides thermal conductivity increasing, because the number of contacts between the yarns are grown.

High-strength and high-modulus carbon fibers are produced in the form of yarns, tows and tapes. The disadvantage of these materials is their high cost. Medium strength carbon fibers are used to produce carbon clothes. Such materials cost is significantly lower than high-strength and high-modulus carbon fibers. At the same time, their mechanical performance fully meet the thermal protection requirements.

The carbon fiber surface has lower wetting ability comparing with glass fibers, and have lower shear stress. The carbon fibers and the polymer matrix adhesion are increased by these procedures: lubricant films removing, a monomer thin layer applying to the fiber surface with following polymerization, fiber surface etching by oxidizer, surface whiskering, etc.

2.2.4　Boron fibers

Boron fibers with high strength and high elasticity modulus were created by chemical vapor deposition in 1959. Boron fibers were applied mainly as reinforcing fibers for the heavily loaded aircraft structural members. Compared to other reinforcing fibers types, boron fibers have high shear stiffness ($G = 180$ GPa). Boron fibers are one of the most effective CM reinforcers.

The manufactual process of boron fibers is based on the boron deposition on a tungsten wire. The mixture of hydrogen H_2 and boron trichloride BCl_3 are used. Boron is deposited on a tungsten wire with a diameter of about 12 μm, that is electric current heated. The wire heating temperature is about 1,350 ℃. The boron deposition rate is about 0.9 kg per week.

As boron is deposited on a tungsten wire, the fiber electrical resistance is changed, therefore, the temperature along the fiber length is different. Two- or multi-stage circuit is applied for electric current for the temperature distribution compensation.

It is necessary to provide an optimal boron deposition rate. If the deposition rate is not equal to optimal, then the fibers strength is reduced. The boron-tungsten fibers tensile strength is higher than 3,000 MPa. Fibers with short length has 6,890 MPa (after the core removing from the fiber). The boron fibers mechanical performance are given in Table 2.8.

Table 2.8　Boron fibers mechanical performance

Country, fiber grade	Density $\rho \cdot 10^{-3}/$ $(kg \cdot m^{-3})$	Diameter $d/\mu m$	Elasticity modulus E_1^+/GPa	Average strength for sample length 10 mm, σ_1^+/GPa	Ultimate strain $\varepsilon/\%$
USA, Avco (B/W)	2.5	98	390—400	3.39	0.85
Japan, Toshiba (B/W)	2.5	97.2	363—386	3.74	1.0
	2.5	96.8	378—388	3.58	0.93
	2.5	99	374—393	3.23	0.84
France, SMPE (B/W)	2.5	100±5	408	3.57	0.88
Germany, Wacker-Chemie (B/W)	2.54	100±5	420	3.1	0.74
Russia (B/W)	2.5	195±3	394	2.95—3.5	0.75—0.9

A fibers strength high level is provided by small boron crystals with sizes of about 2 nm. The boron is deposited in polycrystalline form with the individual crystals dimensions of some hundreds nanometers and more at temperatures above 1,400 ℃. This structure provides boron fibers with a low-level performance.

The boron fibers destruction, as a rule, starts near the surface defects. The fibers surface etching allows to reduce the defects number and to increase the strength performance. The deposit reaction compound purity also influences for the boron fibers strength increasing.

The strength degradation is started with the boron fibers heating and temperature increasing. In the regular atmospheric, the strength degradation becomes significant at 350—400 ℃. The oxide film B_2O_3 on the fiber surface is the reason for the strength degradation, because it decreases the effective fiber cross-section. In argon, the boron fibers strength is kept up about to 600 ℃.

In the United States, in 1990, the boron fiber production was more than 20 tons. For industry, fiber diameters of 100, 140, and 200 microns are produced. Other diameters fibers are produced in small batches.

Boron-tungsten fibers have a high cost, because the tungsten wire is expensive. There is manufacturing process for using cheaper carbon fiber instead of expensive tungsten wire. The boron fibers diameter produced by this way does not exceed 75 microns. The carbon fiber with a pyrolytic graphite layer provides the boron fibers diameter of up to 120 microns.

The boron-carbon fibers elasticity modulus is lower than boron-tungsten elasticity modulus (for fibers with a diameter of 100 μmit is about 360 GPa and 400 GPa, respectively). It depends on the elasticity moduli of carbon yarn and tungsten wire.

The most produced boron fibers are processed into continuous tapes and mats impregnated with an epoxy binder (prepregs). The prepreg is a semi-finished product (blank) that consists of a thermosetting binder with a fibrous filler. Reinforcing fibrous filler in prepregs can be: oriented yarns, tows, tapes, clothes, mats. Epoxy binder is applied to the fiber in two ways:

　　1) by passing fiber through the impregnating rolls;

　　2) by transferring a binder from a special paper.

Special preservatives are applied into the binder in order to increase the prepregs storage time. The prepregs storage time at usual temperature is several weeks, but at − 18 ℃, they can be stored for several years without loss of performance.

The boron fibers have widespread due to their high mechanical performance. They are used for the polymer and metal CM manufacturing. Aluminum is usually used as a metal matrix. Such composites can operate at temperatures up to 640 °K. Shuttle spacecraft structural members were successfully made of boron-aluminum. The boron fibers, high cost is the reason for their limit application.

Boron-fiber reinforced plastic are applied: C - 5a Galaxy super-cargo aircraft — slat; C - 130 cargo aircraft — wing torsion box; F - 15 fighter — tail unit and reinforced floor; etc. Boron fibers are successfully used in the manufacture of sports equipment: tennis rackets, golf crooked sticks, bicycles, light fishing rods, etc.

One of the problems related with the boron fibers application in metal CM is that boron actively reacts with a metal base (for example, aluminum or titanium). In practice, this problem is solved by creating special coatings on the boron fibers surface. Typically, boron carbides (B_4C), silicon (SiC) or boron nitride (BN) are used as such coatings.

2.2.5 Organic fibers

For the CM production, organic aramid and polyethylene fibers are mainly used. Aramid fibers are aromatic polyamide materials. At first they were produced in 1971 under the Kevlar trademark (DuPont company). The Kevlar's advantages are high mechanical performance in a wide temperature range and good chemical resistance.

Aramid yarns are formed by the "dry-wet" method. The raw polymer is pickled at strong acid at about 50—100 ℃ and its purified solution is extruded through draw holes. The single fiber diameter is about 12 μm. Further, the fibers pass through cold water ($t \leqslant 4$ ℃). The resulting yarn is spooled. The yarn reinforcement is before spooling.

Kevlar is a rigid-chain highly oriented polymer. Its chemical structure is shown in Figure 2. 10. Benzene rings provide the polyamide macromolecules orientation along one axis. The macromolecules axes are coincided with the Kevlar fibers axes. The interaction is carried out by hydrogen linkage in the transverse direction. Covalent linkages are in the longitudinal direction. Thus, Kevlar fibers have a high mechanical performance anisotropy. The fiber has a high longitudinal and low lateral strength. The Kevlar fiber structure is similar to a set of long, laterally connected rods.

The aramid fibers chemical structure determines their unique performance. They have the highest strength and elasticity modulus among organic fibers. The density of Kevlar fibers is about 1,440 kg · m^{-3}; the tensile strength is about 2,700—3,300 MPa; the maximum elasticity modulus is 96. 5 GPa (Kevlar - 29) or

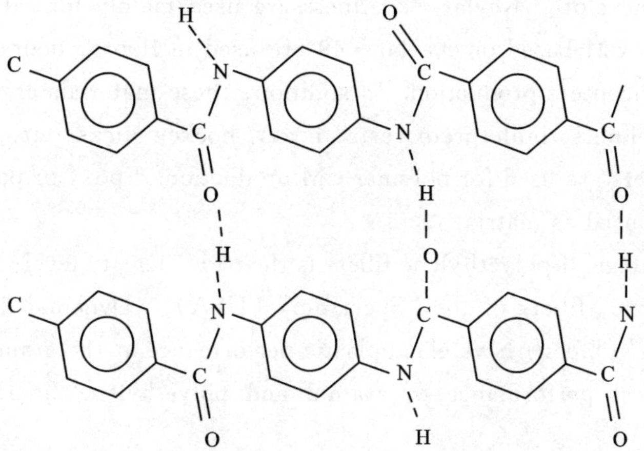

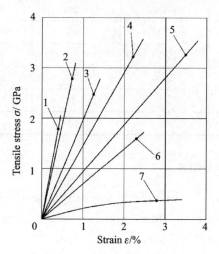

Figure 2.10   Kevlar fiber chemical structure

about 128 GPa (Kevlar – 49). Aramid fibers are more plastic compared to glass and carbon fibers. Their strain under tension is about 2%—4%. The disadvantage of aramid fibers is the moisture absorption, as a result of which their performance are reduced by 15%—20%.

The performance of various types of reinforcing fibers used for CM production is shown in Figure 2.11. The Kevlar – 49 strength in the longitudinal direction are unique even in comparison with inorganic reinforcing fibers.

1—high modulus carbon yarn (E = 414 GPa); 2—boron fibers (E = 379 GPa);

3—high strength carbon yarn (E = 262 GPa); 4—Kevlar-49 fiber roving (E = 131 GPa);

5—S-glass fiber roving (E = 82.7 GPa); 6—E-glass fiber roving (E = 68.9 GPa); 7—nomex fiber

Figure 2.11   Dependences tensile stress-strain for various fibers

DuPont produces such well-known aramid fibers grades as Kevlar – 29 and Kevlar – 49. Kevlar – 29 is technical yarns for various purposes; Kevlar – 49 is yarn, roving and cloth. Kevlar – 29 fibers are used mainly for cables and ropes, cord tires, etc. CM based on Kevlar – 49 are used in floors, doors, bulkheads of aircraft and helicopters production. In addition, these materials are used in rocket engine case, helmets, bulletproof vests, boats, hockey sticks, etc.

Kevlar fibers are used for polymer CM production. Epoxy or polyester binders are most often used as matrix.

The advantage of polyethylene fibers is their low density level: $0.97 \text{ g} \cdot \text{cm}^{-3}$. At the same time, fibers of the "Spectrum" (USA), "Dynema" (Netherlands), and "Tekmilon" (Japan) have elastoplastic performance at the aramid fibers level. Some mechanical performance of aramid and polyethylene fibers are given in Table 2.9.

Table 2.9   Aramid and polyethylene fibers mechanical performance

Grade (country, company)	Density $\rho/(\text{kg} \cdot \text{m}^{-3})$	Tensile stress σ_1^+/MPa	Elasticity modulus E_1^+/MPa	Strain/%
Armos (Russia, VNEEPV)	1,450	5,000—5,500	140—142	3.5—4.5
CBM(Russia, VNEEPV)	1,430	3,800—4,200	120—135	4.0—4.5
Terlon (Russia, VNEEPV)	1,450	3,100	100—150	2—3.5
Kevlar – 29 (USA, DuPont)	1,440	2,920	69—77	3.6
Kevlar – 129 (USA, DuPont)	1,440	3,200	75—98	3.6
Tvaron (Netherlands, AKZO)	1,440	2,800	80—120	3.3—3.5
Technora (Japan, Teidgin)	1,390	3,000—3,400	71—83	4.2
Spectra 900 (USA, Ellayad)	970	2,570	50—120	3—6
Dinema (Netherlands, DCM)	970	up to 3,350	50—120	3—6
Techmilon (Japan, Micui)	960	3,500	100	4

2.2.6 Ceramic fibers

Ceramic fibers have a high-level mechanical performance, high chemical resistance, heat resistance and heat strength, etc. Ceramic fibers are especially useful for the metal matrix composites production.

Ceramic fibers are made from strong, refractory, chemically resistant materials: aluminum and zirconium oxides, boron nitride, boron and silicon carbides, other carbides, nitrides and oxides.

The polycrystalline ceramic fibers creation process includes three stages:

1) a suspension or colloidal solution of an organometallic compound creation;

2) fiber creation by forcing a suspension (solution) through a draw hole or a

thin film creation on a smooth surface;

3) fiber roasting to remove organic substances and a dense structure creation.

The polycrystalline ceramic films creation process includes three stages:

1) raw material creation by a metal salt dissolving in a carboxylic acid (acetic, formic, citric);

2) a solution distribution at the glass surface and its drying at 80 ℃ temperature;

3) strips roasting at the temperature of 1,200 ℃ for organic substances burning.

One of the most promising CM reinforcements are silicon carbide fibers. The silicon carbide fibers density is 3,900—4,050 kg · m^{-3}. The tensile strength is about 3,000—3,400 MPa; the elasticity modulus is about 450 GPa. The fibers strength is depending on the local defects' type and number on their surface (thickenings, foreign particles, sticking points). Silicon carbide fibers are heat resistant and heat strength, and able to withstand high temperatures long time. For metal CM production, it is important. The silicon carbide fibers application in matrix of titanium or heat-resistant alloys is promising. Silicon carbide fibers can be coated with tungsten carbides WC, tantalum TaC, hafnium HfC, titanium TiC. Silicon carbide can also be applied in boron fibers.

The silicon carbide fibers mechanical performance are higher than boron ones. Another important advantage of silicon carbide fibers is their lower cost compared to boron fibers. This fact is due to the raw materials' lower cost, the silicon carbide's higher deposition rate, and other manufactural features.

2.2.7 Crystal whiskers

Whiskers are one of the most interesting types of CM reinforcement. Their main features are single-crystal structure and small size (Figure 2.12). The whiskers' diameter reaches 10 microns. The length to diameter ratio is 20—100, sometimes over 1,000.

The whiskers have a small number of structural defects, so their strength is quite high. Some whiskers properties are given in Table 2.10.

Figure 2.12 α-Al$_2$O$_3$ (sapphire) whiskers with the diameter from 1 μm to 15 μm

Table 2. 10   Whiskers performance

Whiskers materials	Melting point/℃	Density/ (kg · m⁻³)	Elasticity modulus E/GPa	Tensile stress (max)/ GPa	Specific strength/km	Specific elasticity modulus/km
Graphite	3,640	1,660	695	19	1,075	42,500
SiC	2,665	3,320	480	20	650	15,200
BeO	2,560	2,770	340	13	450	12,300
B₄C	2,450	2,490	440	13. 5	550	20,000
Al₂O₃	2,040	3,880	625	27	525	10,700
Si₃N₄	1,675	3,320	300	13. 5	425	12,000
Fe	1,540	6,370	195	13	200	3,200
Cr	1,665	7,500	225	9	125	3,200
Cu	1,080	8,860	125	3	25	1,500

The strength of whiskers with a diameter that less than 2 microns are close to theoretical values. However, the whiskers' diameter is larger and the yield's strength is lower. The ultimate strength dependencies of the aluminum oxide and iron whiskers are shown in Figure 2. 13. With the growing of whisker diameter the surface defect number also grows.

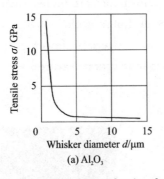

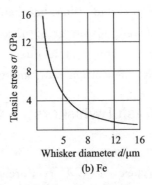

Figure 2. 13   Dependencies of whiskers strength on their diameter

Al₂O₃ and SiC are preferable for the CM reinforcement. They have high-level mechanical properties, relatively low density and are well compatible with metal matrices.

Whiskers are able to nucleate and grow naturally, without human activity. Examples of natural origin crystals are fibers of asbestos, jade, rutile, gold, silver. However, for the CM manufacture, whiskers grown under industrial conditions are used. The whiskers application is determined by crystals high-level mechanical properties.

Currently, there are several methods for whiskers production: growth from coatings; electrolyte deposition; vapor deposition; creation by chemical methods; crystallization from solutions; fiber whiskerization.

1. Growth from coating

For the whiskers' growth from coatings, some fusible metals are used, for example zinc, cadmium, tin, indium. Coatings several microns thick are applied electrolytically, by vapor deposition or by dipping into a molten metal bath. The whiskers' diameter reaches 2 microns (sometimes 5 microns); the length is several millimeters.

2. Electrolyte deposition

Whiskers are growing by the electrolytic deposition method from iron, silver, copper with a diameter of 0. 1—5 microns and a several millimeters length. Organic impurities (gelatin, glucose, oleic acid) and inorganic particles (glass, graphite) are added to the electrolyte. It is needed to limit the crystal growth in the transverse direction and optimize the material deposition process. The strength decreases due to the impuritying contaminate of the crystals.

3. High-pressure arc

Also, whiskers are grown at a high pressure arc in arc discharge chambers with graphite electrodes. A fishhook gradually grows on the negative electrode when a positive graphite electrode evaporates. A several centimeter fishhook is split and whiskers are taking from it. The whiskers length reaches 3 cm; the diameter is up to 5 microns.

4. Vapor deposition

The vapor deposition method produces whiskers of aluminum, barium, tungsten, iron, copper, nickel, platinum and other materials. The whiskers' diameter varies from a micrometer till 40 microns; the length reaches several centimeters. A vacuum furnace or an inert gas furnace are applied in this method. There are zones that differ in temperature level in the furnace. The raw substance heats up, evaporates and transfers (in the gas form) to the colder zone of the furnace, where it condenses in the form of whiskers. The cold zone temperature is 20—200 degrees below the material melting temperature.

5. Chemical methods

Chemical methods for whiskers production are based on the vaporized substance chemical interaction and the environment. The resulting components are transferred to the deposition zone, where the whiskers crystallization takes place.

Chemical methods for monocrystalline fibers production also include the VLS method (vapor —liquid —solid). The crystal formation mechanism is based on: the gaseous material atom condensation in a liquid material particle, which is an intermediate medium; a liquid supersaturation; and the whiskers creation by the material segregation on the carrier.

Silicon, germanium, arsenide, and gallium phosphide whiskers were created by the VLS method. Solvents were gold, silver, platinum, palladium, nickel and copper.

The refractory compound whiskers are created mainly by vapor deposition.

2.3 Reinforcing elements formation

Monofibers, which were created by various methods, can be directly used to CM reinforcement. However, fibers are often transformed to yarns. Yarn is twisted fibers. Yarn can be homogeneous, i. e. , twisted from the same fibers, and mixed, i. e. twisted from different fibers. Yarn can be applied in composites reinforcing in the form of threads, as well as in the form of clothes.

The linear density of fibers or yarns are evaluated by tex unit. Tex is the ratio of the yarn (fiber) mass m to its length L, such as 1 tex $= 1$ g $\cdot$ km$^{-1} = 1$ mg $\cdot$ m^{-1}. In a numbering system, the value of N shows how many meters of yarn are contained in a unit of mass, such as 1 g.

$$N = \frac{L}{m} \qquad\qquad (2.1)$$

The clothes, which can be created during weaving of yarns and fibers, consist of longitudinally arranged elements (warp) and transverse elements (weft). There are: tabby, print, satin, twill and knit weaving with according to the structure (Figure 2. 14). The main clothes specifics are: fiber type, weave type, finishing type, width, thickness, weight per square meter, number of warp and weft per unit length (end count), breaking weight, strain at break.

The cloth width can be narrow (40—75 cm) (tape), medium (75—100 cm), broad (100—150 cm), very wide (150—200 cm and more). If the cloth weight of one square meter is up to 100 g, then it is light, for 100—500 g $\cdot$ m^{-2} it is medium, and more than 500 g $\cdot$ m^{-2} it is heavy.

The structure of tabby, satin and twill clothes are shown in Figure 2. 5. Performance of some organic-cloth are presented in Table 2. 11.

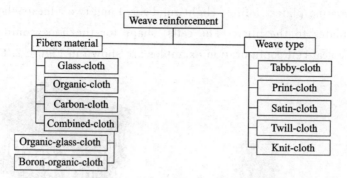

Figure 2.14　Weave reinforcement classification

Table 2.11　Organic-cloth performance

Material grade (Country)	Weave type	Aerial density/ $(kg \cdot m^{-3})$	Thickness/ mm	End count/ $(tex \cdot mm^{-1})$ warp	weft	Average strength/ GPa warp	weft	Tensile strain $\varepsilon/\%$ warp	weft
SVM High-strength CM (Russia)	Tabby	—	0.45	142	142	0.39	0.39	—	—
		0.11	0.25—0.3	44.1	47	0.24	0.27	14	12
		0.075	0.15	26.5	30	0.28	0.35	10	9
	Basket 2/2	0.18	0.35	59	74	0.27	0.31	9	11
		0.11	0.20	43	44	0.26	0.26	10	10
	Satin 8/3	0.16	0.40	75	69	0.26	0.21	12	9
	Uniaxial tape	0.17	0.35	168	25.7	71	—	7.5	—
Kevlar-49 (USA)	Tabby	—	0.45	140	130	—	—	—	—

The fibers (yarns) with high plasticity should be applied in such CM and it is the disadvantage of CM that created by weaving. Metal wires have especially bad performance. At the warp and weft contact points, wires are deformed, and defects are started, which reduces the CM performance.

The knitted nets provide more wide performance for CM. The stockinette and rib structures are shown in Figure 2.15. The high-strength metal wire with a diameter of 0.02—0.2 mm and with a relatively low strain 1%—3% can be applied in the knitted nets manufacture. Combined various materials and fiber construction nets are produced by knitted processing.

Reinforcing fibers can also be combined into non-woven materials (such as felt), in which their regular orientation is absent. Methods of liquid, air, gravity, vacuum felting have been developed.

The sandwich (honeycomb) CM are special structural type. They are applied in almost all modern types of civil and military airplanes and helicopters. The honeycomb structure is shown in Figure 2.16 as an example. It consists of two strong overcoating plates, a filler (light rigid core) and two adhesive layers that are linking the plates to the filler. The cells' shape for the honeycomb core can be different. Typical cell configuration examples are shown in Figure 2.17.

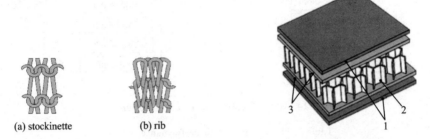

(a) stockinette   (b) rib

Figure 2.15   Knitted nets weave types

1—overcoating plates; 2—filler (core); 3—adhesive layers

Figure 2.16   Honeycomb structure

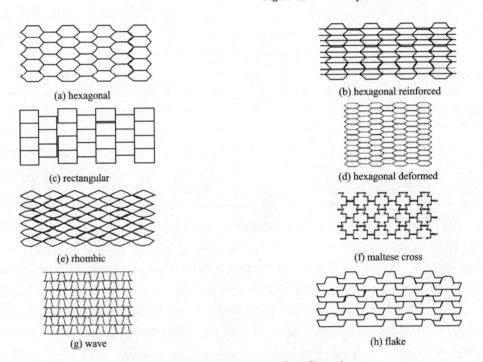

(a) hexagonal

(b) hexagonal reinforced

(c) rectangular

(d) hexagonal deformed

(e) rhombic

(f) maltese cross

(g) wave

(h) flake

Figure 2.17   Honeycomb core cell configurations

A honeycomb core increases the structural flexural stiffness. The honeycomb efficiency can be illustrated by the data which is given in Table 2.12. This table is based on various thicknesses aluminum sheets structures (Figure 2.18). For example, the relative difference of density by 6% are variants (a) and (c) (Figure 2.18), however, their stiffness difference is by 39 times.

Table 2.12   Honeycomb performance from aluminum and different thickness

Honeycomb performance	Structure type		
	Figure 2.18(a)	Figure 2.18(b)	Figure 2.18(c)
Buckling stiffness/$(kg \cdot cm \cdot cm^{-1})$	1,822	13,933	71,004
Specific stiffness changes	1.0	7.4	39.0
Density/$(kg \cdot m^{-3})$	29	30	31
Specific density changes	1.0	1.03	1.06

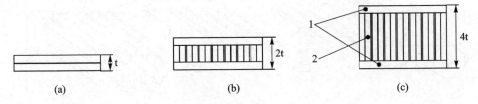

(a)   (b)   (c)

1 —overcoating plates; 2 —core

Figure 2.18   Honeycomb structure with different core thickness

The overcoating plates materials for honeycomb structures can be fiberglass prepregs, carbon fibers prepregs, aluminum, titanium alloys, steel, etc. The honeycomb core's main functions are to provide the common structural stiffness and to support the shear loads. The core must be rigid and light for successful providing these functions. Honeycomb core materials can be wood, metal, kraft paper, aramid paper, foams (polystyrene foam, foamed metal). Foam-core structures have been used about 40 years. The efficiency of fiberglass honeycomb structure is confirmed by their application in the Gemini and Apollo spacecraft.

Two manufactural processes are applied in the honeycomb core production: lay-up pulling and corrugating (Figure 2.19).

The lay-up pulling is shown in Figure 2.19 (a). The sheets are collected to the lay-up. They are joined by gluing, welding or soldering along base lines. After that, the joined lay-up is pulled and formed a cell structure.

At the second process of the first stage, the raw sheet (fiber-plastic, paper, metal, etc.) is corrugating. After that, corrugated sheets are joined (welding, gluing or soldering) [Figure 2.19 (b)].

For aluminum alloys, the honeycomb effective density created by the pulling is $32—192$ kg $\cdot$ m^{-3}, and the honeycomb effective density created by the corrugating is $128—880$ kg $\cdot$ m^{-3}.

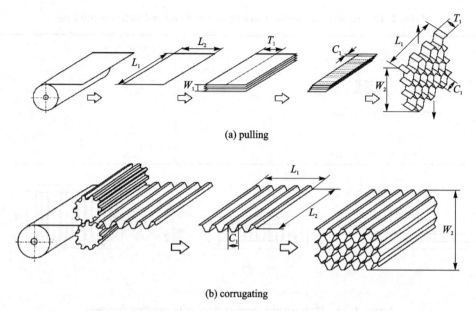

(a) pulling

(b) corrugating

Figure 2. 19 Manufactural processes for honeycomb core creation

Questions

(1) List the types of metal matrices.

(2) Briefly describe the features of metal matrices.

(3) List the types of polymer matrices.

(4) Briefly describe the features of polymer matrices.

(5) List the types of reinforcing elements.

(6) Briefly describe the features of metallic fibers.

(7) Briefly describe the features of glass and silica fibers.

(8) Briefly describe the features of carbon fibers.

(9) Briefly describe the features of boron fibers.

(10) Briefly describe the features of crystal whiskers.

(11) Explain the weave reinforcement classification.

(12) List the types of honeycomb core cell configurations.

Chapter 3   Production of Polymeric Composite Materials and Products from Them

The most important feature of the modern CM production is that, as a rule, the composite and its manufactural processing are developed for a particular structure. In this regard, the structure, of the CM for the manufactural processing development is three parts of a single task. Given above, the new CM's developing and their manufactural processing should be considered together and the part's purpose, its shape, dimensions and operating conditions should be taken into account. Science materials, manufactural and structural aspects of the tasks related with the CM parts creation are simultaneously discussed here.

In the general case, the polymer CM production includes the following operations:

1) reinforcing filler preparation;

2) polymer matrix preparation;

3) matrix and reinforcement coupling;

4) parts forming;

5) matrix polymerization;

6) part machining (if necessary);

7) part monitoring.

Besides, the main attention will be paid to the processes of reinforcement preparation, the matrix and the reinforcement coupling, and parts forming for polymer CM.

Currently, a large number of methods have been developed for the polymer CM manufacture. The main forming methods include:

1) impression molding;

2) flexible diaphragm molding;

3) pressure molding;

4) compression forming;

5) winding;

6) pultrusion;

7) pre-molding of blanks and mats.

These processes implementation features and their varieties are given in Table 3.1.

Table 3.1   Main processes of CM parts forming

Process	Components	
	Reinforcement	Matrix
Impression molding: 1) hand lay-up 2) schoop 3) automated placement	Short fibers, tapes, clothes Mats, tapes	Polyester, epoxy, phenolic, furan, thermoplastic
Flexible diaphragm molding: 1) vacuum molding 2) vacuum-autoclave molding 3) vacuum-press-chamber	Clothes, mats, tapes	Epoxy, polyester, polyamide, phenolic, polysulfone
Pressure molding: plenum-chamber process	Clothes, mats, tapes, short fibers	Polyester, epoxy, phenolic, furan, polyamide
Compression forming: 1) compression molding 2) injection 3) termocompression	Clothes, mats, tapes Short fibers, clothes, mats, tapes	Polyester, epoxy, phenolic, silicone, thermoplastic
Winding: 1) wet 2) dry	Fibers, types, yarns, clothes, films Prepregs, fibers, types, yarns, clothes	Epoxy, polyester, phenolic, polyimide, thermoplastic
Pultrusion	Fibers, yarns, woven tape, woven blanks	Thermosetting, thermoplastic
Pre-molding of blanks and mats	Short chopped fibers, whiskers	Aqueous solutions of ethyl alcohol, carbon tetrachloride, polymer binders

The methods applicability is determined by many factors, including the parts shape and dimensions, their quantity, process performance, process parameters, part's requirements (strength, accuracy, etc.).

3.1 Impression molding

The impression molding is widely used for the polymer CM production. In this way, boats hulls, car parts, furniture, etc. are produced. The molding process consists of layer-by-layer laying of reinforcement on the prepared mold, their liquid polymer molding and subsequent compaction for the air bubbles removing. The liquid polymer molding can be carried out not only after its laying in the mold but also at the previous stage. There are three ways of impression molding method:

1) hand lay-up;

2) automated placement;

3) schoop.

3.1.1   Hand lay-up

The main type of equipment used in this method is the mold. There are positive and negative dies. The positive die provides a smooth and accurate inner surface of the part. The negative die provides a high quality of the external part surface. The polymer CM structure produced by the negative die is shown in Figure 3.1. The example of a polymer CM produced by the positive die is shown in Figure 3.2.

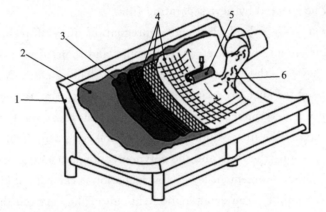

1—die; 2—film; 3—gel coat; 4—glass fibers; 5—hand roller; 6—resin with polymerization accelerator mixt

Figure 3.1   Hand lay-up for negative die

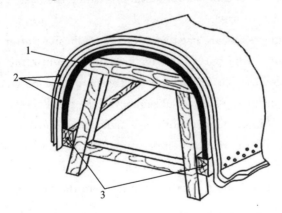

1—die; 2—glass fibers layers; 3—glass fibers layers' attachment

Figure 3.2   Hand lay-up for positive die

The main elements of the hand lay-up process are the die, the separating film, the gel coat, some glass fibers layers, a hand tools and resin with polymerization accelerator mixt. The die surface must meet the requirements of geometry, rigidity, adhesion, etc. Such materials as wood, plywood, metal, concrete, gypsum, fiberglass, etc. can be used to make the die.

As the bonding adhesive does not adhere to the die surface, special separating films are used. However, if the bonding adhesive gets to the die surface, the composite will adhere to the die. Adhesion will lead to part surface defects and the die damage itself. This is especially true for die created of porous materials, such as wood, gypsum, etc. Therefore, the die surface is covered with paraffin, wax or other films that prevent the die adhesion with the composite. The die can be used repeatedly. In order to provide the highest part surface quality, it is needed to polish the die after each molding cycle. After several parts' production the die surface should be covered by new separating film.

On the next stage, the laying and impregnation of reinforcement layers are present. Reinforcement sheets are cut from rolls, and then, if necessary, with the help of special devices, they are cut according to the shaping plate. Mats and woven roving can be used for such process. The layer's number of mats and roving, their location sequence is determined by the required part thickness and the requirements that it should have. The fiber volume fraction in the composite is 35%—50%.

For impregnation, the resin and the polymerization accelerator are mixed in the required ratio. The reinforcement layers impregnation is carried out either in the die or outside it. Air bubbles and delamination should not be between the layers. The layers are compacted with brushes, rubber and gear rollers in order to prevent such defects. In the USA, this compaction method is called the basket and brush method because simple tools are used.

If thermosetting resins are used as a bonding adhesive, after the completion of the laying operations, the part must be kept in die until the matrix hardening procedure is completed. The application of thermoplastic polymeric materials has some features. In particular, each subsequent layer can be sequentially molded to the previous one. Molding operations are performed until the part reaches the required thickness. At the regular temperature, the polymers will not be adhesive, therefore, for bonding each layer to the previous one, it is necessary to use a special heating tool that provides a temperature higher than the melting temperature of the bonding adhesive.

3.1.2 Automated placement

The automated placement process is based on the laying machines application that capable of moving along complex trajectories. This process is implemented for the complex shape open shells creation. Machine control is carried out according to programs developed for each specific part. The prepreg is the base material for automated placement process. After laying out the prepreg layer in order to eliminate air bubbles, it is compacted by rolling with rollers. The automated placement device scheme of a double-curvature shell prepreg is shown in Figure 3. 3.

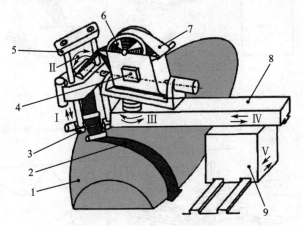

1—die; 2—prepreg and separating film tape; 3—head; 4—heater; 5—frame;

6—prepreg spool; 7—separating foil; 8—lathe; 9—lathe carriage

I , II , III , IV , V —machine parts mobility directions

Figure 3. 3　Automated placement process scheme

When laying prepreg on the surface of the mandrel of complex shape, the formation of folds is possible. In order to reduce the risk of folding, a small prepreg or a woven plain weave prepreg can be used. With a certain width of the prepreg, it can be laid without the formation of defects on gently sloping surfaces of double curvature. Warming up the prepreg increases its ability to deform in the plane of reinforcement.

3.1.3　Schoop

The common schoop process of polymer CM is shown in Figure 3. 4. Shredded fibers obtained by roving using a chopper and resin are simultaneously fed into the mold.

In fact, the schoop process can be considered as a variant of the manual lay-up of parts.

When parts manufactured in large batches, it is economically viable to use automated station to perform schoop process. The automation station application meets the labor safety requirements. In Figure 3. 5 it is shown one type of such automation station for the polymer CM schoop process.

The automation station for the polymer CM schoop process consists of: pumps

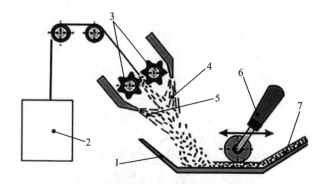

1—mold; 2—roving; 3—chopper; 4—resin with accelerator;
5—resin with catalyst; 6—roller; 7—compacted layer

Figure 3. 4   Dual tank airless schoop station and schoop process

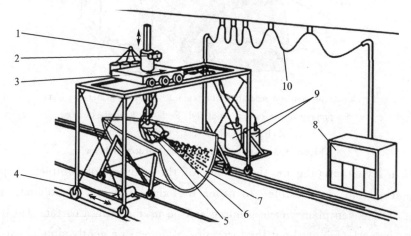

1—fibers feeding; 2—vertical driver; 3—cross driver; 4—longitudinal driver; 5—swinging head;
6—sputtering device; 7—mold; 8—controller box; 9—resin feeding; 10—cable

Figure 3. 5   Three-axial automation station for the polymer CM schoop process

for resin and catalyst; catalyst pressure tanks; solvents injection tanks; grinding reinforcement choppers; air and materials feeding controllers; measuring tools; feeding pipes; sputtering devices.

An important part of such equipment is a device. There are various types of sputtering devices. The most typical sputtering devices (Figure 3. 6) are:

1) with external mixing;

2) with airless external mixing;

3) with internal mixing;

4) with airless internal mixing;

5) with twin-feed.

In the sputtering device with external mixing there are four nozzles [Figure 3. 6 (a)]. Resin is fed through two nozzles, and a catalyst sprayed with the air through the rest. The chopped fiber is fed through the hole in the central zone. The fiber is coated with resin and catalyst and deposited on the mold surface.

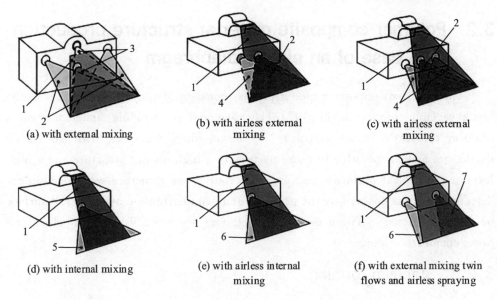

(a) with external mixing

(b) with airless external mixing

(c) with airless external mixing

(d) with internal mixing

(e) with airless internal mixing

(f) with external mixing twin flows and airless spraying

1—grinded fibers; 2—catalyst; 3—polyester; 4—resin; 5—mixture of resin, catalyst, air; 6—mixture of resin, catalyst; 7—mixture of resin, accelerator

Figure 3.6   Typical sputtering devices

Sputtering devices operated on the airless external mixing principle are shown schematically in Figure 3.6(b) and (c). The catalyst stream is fed through nozzles at a pressure sufficient to spray without air.

Often, air-internal mixing devices are used as sputtering devices. Before resin injection into the mold, a mixture of resin, catalyst and air done [Figure 3.6 (d)]. The chopped fiber is fed into the upper part of the mixing fan, coated with a binder and applied to the mold surface.

The catalyst and resin under pressure are mixed in the mixing chamber and fed to the nozzle under high pressure ($\geqslant$ 5.5 MPa) in sputtering devices with airless internal mixing. This pressure is sufficient to spray without using additional air flow.

To obtain polymer CM by schoop process, airless systems with twin channels are also used. These systems have two nozzles through which a resin and catalyst mixture, and a resin and accelerator mixture are fed. The streams are mixed at a distance of about 150 mm from the nozzles. The chopped fibers stream is mixed with them at the same time. This composition is applied to the mold surface. The operational pressure of such devices is about 2.7—7.5 MPa.

It is important to obtain a high-quality polyester outer resin layer (gelcoat) for the schoop process. The part surface quality depends on the workers skills. Surface inspection is only after the full drying and removing the part from the mold.

3.2 Polymer composite material structure production on a base of an elastic diaphragm

The elastic diaphragm molding process consists of the following stages: the blank vacuum compaction in the rubber bag and its molding under the gases uniform pressure (in an autoclave). The molding processes with an elastic diaphragm make it possible to make parts with a high-quality structure and a high level of mechanical performance. The materials dense structure with a minimum defects number is ensured by the pressure uniform distribution over the full surface of the molded part. There are some methods: vacuum, vacuum-autoclave and press-chamber molding.

3.2.1 Vacuum forming

Devices' schemes that applied for the vacuum forming with an elastic diaphragm are shown in Figure 3.7. The scheme of Figure 3.7(a) shows the positive die application, the scheme of Figure 3.7(b) shows the negative die application. The main equipment elements are the base (1), the mold (3), the formed blank (6), the couling plate (4) (or cellophane), the rubber bag (5) and holders (2). The couling plate is a thin skin (its thickness is about 0.5—1.5 mm), the shape of which corresponds to the molded part shape. The couling plate is made of aluminum alloy, fabric-reinforced laminate or fiberglass laminate. It compacts the blank from all sides in time of pumping air from the rubber bag.

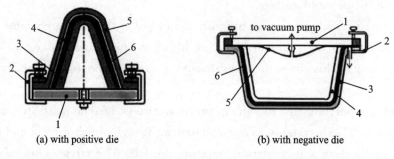

(a) with positive die   (b) with negative die

1—base; 2—holder; 3—mold; 4—couling plate; 5—rubber bag; 6—blank

Figure 3.7　Equipment scheme for the vacuum forming process

The external pressure compacts the reinforcement layers, ensures high-quality deep fiber impregnation with a binder, removes air bubbles, volatile by-products, as well as excess resin from the layers.

The gaseous by-product minimum quantity during the part polymerization is an important requirement for the binder. These by-products create back pressure under the rubber bag and they are the reason of the CM delamination. Molds are made from various materials, including metal alloys, gypsum, laminate, wood, etc. Wooden molds are used for small parts batches with using no-bake binders. The mold material choice depends on its thermal linear expansion coefficient. Then mold dimensions are higher during its heating, then the part accuracy is lower.

The molds' surface is covered with an antiadherent layer in order to prevent the moldable material seizing with the mold. The paste wax (up to 121 ℃), the fluorocarbon (up to 177 ℃), a silicone lubricant (up to 204 ℃), a silane resin lubricant (up to 482 ℃) can be used as antiadherents. The antiadherent layer is renewed after 4 or 5 times. The antiadherent functions can be performed by a cellophane film.

Clothes or mats that previously impregnated with a binder are piled over the antiadherent layer in the amount required. The layered clothes are rolled onto the mold in order to remove gases bubbles between layers, increase density. Separating drain films (clothes) are piled over the blank surface. These films absorb the bleedout in the material hardening process. Sometimes couling plate is used as an antiadherent. The top layer that is near the vacuum rubber bag prevents the bag from seizing to the blank.

Special rubber grades that can be used for resins and high temperatures, rubberized fabrics, dimethyl silicone that reinforced with fiberglass can be used as materials of a vacuum bag (elastic diaphragm). It is possible to use heat-resistant films, for example, polyimide (up to 316 ℃), polyamide (up to 230 ℃), nylon (up to 204 ℃), polyvinyl alcohol films (up to 121 ℃). Pipe connectors fused or glued into the elastic diaphragm connect it to a vacuum pump. The number of pipe connectors depend on the diaphragm area.

An important manufactural requirement is the elastic diaphragm tightness.

3.2.2   Vacuum-autoclave forming

The vacuum-autoclave molding process is fundamentally similar to the vacuum forming. However, for its implementation, in addition to the vacuum system, an autoclave is required, which provides additional pressure on the molded part surface.

Autoclaves schematic diagram that is used for polymer CM forming are shown in Figure 3.8. An autoclave is a container in which a molded part is placed. The autoclave includes the following systems: the vacuum providing, the gas supply, the required temperature heating, pressure regulation, emergency depressuring. The autoclave box is protected by thermal insulation. It is possible to process parts with various dimensions. The operational pressure is about 1. 6—3. 0 MPa, and the operational temperature is 250—450 ℃.

　　A part placed inside an autoclave is vacuum compacted by similar process with an elastic diaphragm. Air is pumped out of the autoclave chamber, after that it is filled with a pressurized gas. Overpressure is necessary for the part uniform forming under the elastic diaphragm. At the next stage, the chamber is heated. The medium temperature is controlled by thermocouple-sensing elements. Upon part forming completion and cooling up to 60—70 ℃, the pressure in the autoclave chamber decreases. The autoclave is cooled by a water heat exchanger.

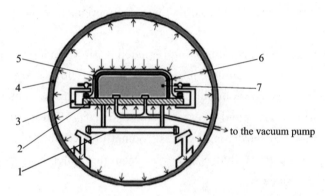

1—bogey; 2—base plate; 3—holder; 4—autoclave box; 5—rubber bag; 6—blank; 7—mold

Figure 3. 8　An autoclave schematic diagram

3.2.3　Press-chamber molding

　　A diagram explaining the press-chamber molding features for polymer CM parts is shown in Figure 3. 9. A mold is used as manufactural equipment, on which

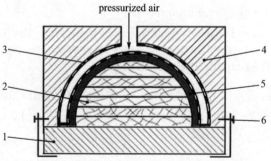

1—base plate; 2—mold; 3—rubber cover; 4—couling plate; 5—blank; 6—holder

Figure 3. 9　Press-chamber molding process diagram

the pressed blank is placed. The blank outer surface is covered with a rubber cover.

When air is supplied, the rubber cover (bag) is stretched and pressed on one side to the blank outer surface, and on the other side to the couling plate inner surface. The rubber cover pressure is the uniform distributed over the total blank surface that provides a good quality of the resulting CM part. Pressing pressure usually does not exceed 0. 5 MPa.

3.3   Polymer composite material structure production on a base of a pressure-forming technology

Two types of manufactural processes are used, which based on the parts forming by pressure: impregnation under pressure and impregnation in vacuum. The forming parts by pressure process provides high quality of the parts (high geometric accuracy, uniform density, absence of voids and delaminations). Aircraft unit's fairings are a typical example of parts obtained by this process.

3.3.1   Plenum-chamber process

Plenum-chamber process for polymer CM parts requires a mold and a tank with a binder, as well as a compressor that pumps the binder into the mold. The equipment scheme is shown in Figure 3. 10. A lamination that is a CM reinforcement is laid up in a mold and closed with a punch. Previously, the mold and the punch surfaces are covered with an antiadherent layer. The blank drying in the form is carried out by blowing through it hot air heated by a heater.

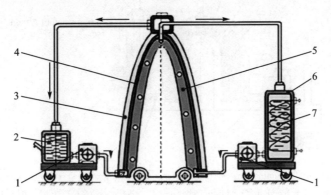

1—compressor; 2—tank with a binder; 3—punch; 4—lamination; 5—mold; 6—heater; 7—organic coolant

Figure 3. 10   Equipment scheme for the plenum-chamber process

A binder is pumped under pressure from a tank into the space between the punch and the mold by using a compressor. The binder gradually flows up and displaces the air from the lamination. The optimal speed of the binder flow up in the mold determines the process performance and the part quality. If the flow up speed is higher than optimal then air bubbles will remain in the binder, adversely affecting the part properties.

　　The bleedout is returned into the tank and applied to reprocessed for other parts production. After mold filling with a binder, the pumping process is stopped and the binder starts to harden. The heating can accelerate this process.

　　Binder and mold temperatures are important manufactural parameters of the impregnation process. The mold temperature control is carried out using a coolant pumped by the compressor through the mold. The part hardening can be carried out by placing the mold in the oven.

3.3.2　Impregnation in vacuum process

　　For vacuum impregnation, molds are similar to those which are used for pressure impregnation. They also have a bonding matrix and a punch (Figure 3.11). Before assembling the mold, a reinforcement lamination is laid up in the bonding matrix. A process specific is that the lamination is impregnated with a binder when pumping out air from the mold. Atmospheric pressure presses on the binder located in the tank and forces it to flow up in the mold and binder is filling the reinforcement. The bonding matrix and punch must be sufficiently rigid in order to prevent their deformation under atmospheric pressure. Otherwise, the binder will not be able to flow up in the mold and uniformly fill its volume.

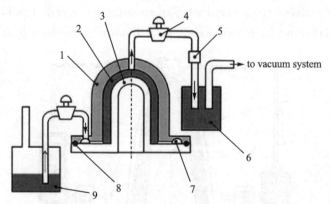

1—punch; 2—lamination; 3—bonding matrix; 4—stop-cock; 5—observing cover;
6—tank with a bleedout; 7—blinder channel; 8—elastic gasket; 9—tank with a blinder

Figure 3.11　Vacuum impregnation process scheme

3.4   Polymer composite material production on a base of compression molding

Compression molding process is widely used for the polymer CM parts manufacturing. About 50% of parts from reinforced plastics are produced by this process. The using of molds provides the require parts configuration that is made of CM.

This process provides high advantages for parts of large batches manufacturing. However, in some cases, the compression molding can be used even for the small-scale production, which is explained by the high quality and by the high accuracy of the size and shape of the parts.

The mold is the main component of any processes that are based on this approach. The mold has operational volume that is corresponded to the shape of the required part. The mold consists of a matrix and a punch. When they close, the working clearance remains in the mold, the thickness of which corresponds to the part thickness.

There are three main processes: direct, injection and thermocompression. The process choice is determined by the mold structure, the CM material performance, and the mold filling way.

3.4.1   Direct compression molding

The material for this process can be: fiber impregnated with a binder; molding polymer grains; binder-impregnated mats or clothes; polymer CM preforms; premix molding compounds. The reinforcement content in the CM obtained by direct compression molding can be 20%—50%, it depends on the processing mods and impregnation ways.

A diagram of the direct compression molding implementation is shown in Figure 3.12. The basic substance is placed into a matrix that heated to a certain temperature. A punch pushes on it. The heated basic substance takes required shape under the applied pressure.

Hydraulic presses are commonly used for direct compression molding process.

The part extraction is carried out after CM curing. The thermoset material curing occurs for some time at a certain temperature. Thermoplastic material cures when temperature drops.

Molds used for the polymer CM manufacture are classified according to various criteria. There are flash, semi-closed and closed molds.

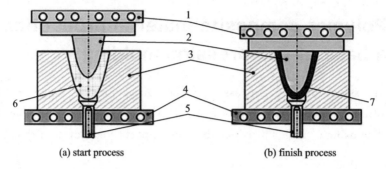

(a) start process　　　　　　　　　(b) finish process

1—heater; 2—punch; 3—matrix; 4—heater; 5—ejector; 6—basic substance; 7—result part

Figure 3. 12　Direct compression molding diagram

Parts made in flash molds have not height accuracy. Flash molds are applied for the polymer CM parts with simple shape and low height of vertical structural elements. The basic substance excess should be about 10%—15% for molding in a flash mold.

Compared to flash molds, semi-closed molds provide a higher material compaction. This makes it possible to create parts with more complex shape in semi-closed molds. The material loss is less in semi-closed molds.

The closed molds' working space volume is similar with the molded part volume. There is no material waste for such molds, because the material does not flow out from them. The closed molds advantages are a material higher compaction, mechanical properties of high level. The closed-type molds' disadvantages are their manufacture complexity and relatively low service life.

Polyvinyl alcohol aqueous solutions, cellophane, lavsan and fluoroplastic are used as anti-adhesive materials for the direct compressing molds. Wax, paraffin, organosilicon lubricants are also applied.

Temperature and time are also important manufactural parameters of the direct molding process. Material heating is necessary to ensure its plasticity and to ease filling the mold working space. When thermosetting binders are used as a matrix material, heating is required to material cure. Molds heating is carried out by using special heaters.

The direct compression method is used to process both thermoplastic and thermosetting materials. More often, this method is used to manufacture parts from thermosetting materials. The thermoplastic materials processing by this method has relatively long duration. The polymer CM with a thermoplastic matrix production requires for alternately heating and cooling the mold in each press cycle. Approximate modes of parts direct compression from various materials are shown in Table 3. 2.

Table 3. 2   Parts direct compression approximate modes

Materials types	Temperature/K	Specific pressure from min to max. /MPa	Curing time/min	Shrinkage capacity/%
Thermoplastic CM	393—523	2—15	5—15	1—3
Elastomers	343—473	0. 4—25	3—60	1—2
Thermosetting CM	393—473	3. 0—75	3—15	0. 2—1. 2
Premixes	393—473	3. 5—7. 0	1—3	0. 5—1. 0

3.4.2   Injection compression molding

This method features for obtaining CM parts are shown in Figure 3. 13.

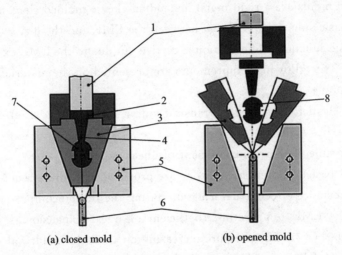

(a) closed mold   (b) opened mold

1—transfer plunger; 2—transfer chamber; 3—wedge-shaped mold; 4—gating channel;
5—cavity retainer; 6—ejector; 7—basic substance; 8—result part

Figure 3. 13   Injection compression molding diagram

The basic substance is placed in the transfer chamber. The material heats up, softens and under the injection punch action through the gating channel enters the mold cavity, which has the parts shape. The material enters into the mold cavity already in a plastic state, which does not allow to create a significant pressure effect on the molds structural members. This fact allows the using of relatively low-strength molds components. The part is kept in a heated state for some time. It is necessary for the material to cure. The mold is opened and the part, together with the gating excess, is extracted from it after material curing. Subsequently, the gating excess is cut off.

The injection molding method has the following advantages:

1) Due to the low temperature difference across the part walls thickness, the

internal stresses are less than with direct compression.

2) The duration of the injection molding process is also shorter compared to direct compression.

3) Injection molded parts have high dimensional accuracy.

4) The mold cavity is closed even before it is filled with material, so there are small flashes numbers on the part surfaces.

The disadvantage is the increased material consumption that elated with a gating channel and an excess.

3.4.3 Thermocompression molding

Special molds are used for the thermocompression molding process. The main parts of such molds are a rigid metal die and an elastic molded element (EME). A compacted basic substance is placed between the EME and the die.

The basic substance deformation is carried out due to the EME expansion when it is heated. The following requirements are presented on the materials from which EME are made:

1) high elasticity required to transmit uniform pressure in all directions;

2) properties stability;

3) high value of thermal coefficient of linear expansion.

Rubbers based on dimethyl silicone are provided with these requirements.

Basic substance effective deformation during the thermocompression process is possible if the EME temperature coefficient of linear expansion is in the range $(250-500)10^{-6} °C^{-1}$. The pressure increasing on the deformable substance occurs during the tool heating. The pressure value is proportional to the temperature change:

$$P = k_P \Delta T, \tag{3.1}$$

where k_P is the coefficient of thermal compression of the EME, which characterizes the pressure increasing inside the closed cave when EME is heated by 1 °C, for rubbers $k_P = 0.5-0.7$ MPa $\cdot °C^{-1}$; ΔT is the difference between the current temperature and the temperature at which the gap between the EME and the mold disappears.

In manufacturing, two molds types are used for thermocompression process (Figure 3.14): molds with constant (a) and variable (b) molding volumes. In the first molds type, the metal punch and the metal die are rigidly connected to each other. The compression pressure in them can be changed by the gap size adjusting and the EME heating temperature. In molds with variable molding volume, the metal punch is connected to the die by means of elastic calibrated elements. In this tool type, the molded pressure is kept at a required level when the pressure is increased in the die due to the punch movement.

The thermocompression process provides the polymer CM parts production with equal thickness. The part produced by this process has a good structure and a high level of mechanical performance. An important advantage of this process is the relatively low labor intensity of tool manufacturing.

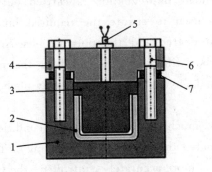 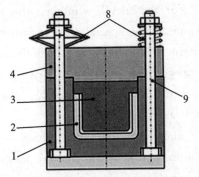

(a) mold with constant molding volume   (b) mold with variable molding volume

1—metallic die; 2—basic substance; 3—EME; 4—metallic punch; 5—pressure sensor;
6—fastener; 7—support bar; 8—elastic calibrated elements; 9—guide pin

Figure 3.14 Thermocompression molding tools diagram

3.5 Polymer composite material parts production on a base of a winding process

The winding process is widely used to obtain parts from polymer CM. For its using, mandrels are needed, the shape of which correspond to the result of the part shape. The winding process is usually used for revolutional shape parts. Examples of such typical parts: pressure cylinders, tanks, rocket engines boxes, aircraft fuselages, etc. The winding process reinforcements are threads, tapes, binders, clothes, thin films. The reinforcement is impregnated with a polymer binder and wounded on a pre-prepared mandrel according to a certain pattern. After the winding process of completion, the binder cures. The reinforcement optimum content in polymer CM produced by the winding process is about 50%—70%. The mandrel either becomes a structural member of the result of a part or is removed.

3.5.1 Winding process types

An important advantage of CM produced by the winding process is the ability of effectively adjusting mechanical performance by structure strengthening in the most loaded directions. This process allows to realize the fibers tensile strength properties to the greatest extent. There are a large number of automated machine tools to perform the winding process. These machine tools are relatively easy to adjust for the various parts production with different complex shapes.

In production, the following winding processes are used: "dry", "wet" and hybrid.

For the "dry" winding, the reinforcement impregnation is carried out in advance. The special impregnating tool is used to ensure the required binder content in the prepregs, the impregnation degree and the polymer viscosity. Prepregs based on glass, organic and carbon fibers are produced in the pre-impregnation process. Epoxy, phenol-formaldehyde, polyimide, epoxy-phenolic resins are used as binders.

The "dry" winding process provides much better workplace conditions (equipment and workplace are less contaminated than with "wet" winding). The volume fraction of reinforcement and binder is more accurately controlled; the CM of mechanical performance are more stable for the "dry" winding process. Due to the prepreg adhesion to the mandrel surface, it is possible to carry out "dry" winding on mandrel with steep grade. The "dry" winding process diagram is shown in Figure 3. 15. The disadvantages of the "dry" winding process are its higher cost compared to "wet" winding (about 2 or 3 times) and the need to use special manufactural equipment.

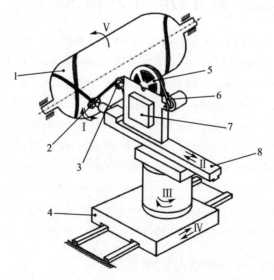

1—mandrel; 2—head; 3—thread tensioner; 4—carrier; 5—prepregs spool; 6—seperating foil; 7—heater; 8—cross slide

Figure 3. 15 "Dry" winding process machine tool

The "wet" winding process involves reinforcement impregnating just before placing it on the mandrel. So, for example, when threads are used as reinforcement, they are unwound from a spool, fed to individual tensioners, formed into strands, and fed into a dipping bath with a binder. Then the impregnated strands are quetched, tensed with the required force and wound on the mandrel that has the part shape. The strand winding onto the mandrel is carried out by a lay-up tool. The "wet" winding process is shown schematically in Figure 3.16. "Wet" winding is used, as a rule, for the large-sized with a complex shape envelops manufacturing.

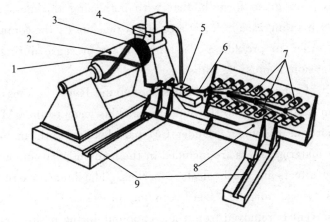

1—twisted layer; 2—mandrel; 3—hoop layer; 4—rotary actuator;

5—dipping bath with a binder; 6—strand; 7—spools; 8—spool holder; 9—guiders

Figure 3.16   "Wet" winding process machine tool

The advantage of this process is lower contact pressure. Its disadvantages are the relatively low winding speed due to the reinforcement impregnation limited speed; uneven binder distribution across the molded part thickness; increased (in comparison with the "dry" winding) amount of binder waste; the inability to use thermoplastics and other binders with high viscosity; contamination of the workplace and tool machines with a binder.

The third process for polymer CM part creation is to pre-wind the reinforcement onto a mandrel and then impregnate it with a binder. Reinforcement impregnation on the mandrel is carried out by binder spraying with a spray gun, brush (manually), using vacuum or pressure tools.

The using of one or another type of winding process depends on the part purpose and its structural specifics, dimensions, weight, type of CM and other features.

3.5.2   Contact pressure for winding process

For the winding process, an important manufactural parameter is the contact pressure that is created by the reinforcement tension or by the pressure rollers. When the reinforcement is wound onto a cylindrical mandrel of radius R with tension T, a radial contact pressure N occurs. This pressure presses the reinforcement to the mandrel surface and compacts the result material. The value of N is determined by the equation

$$N = \frac{T}{R} \sin^2 \alpha, \qquad (3.2)$$

where α is the angle of the reinforcement winding.

It follows from given formula that with increasing thickness of the wound layer, i. e. , with an increase in R, at a constant tension T, the contact pressure N decreases. In order to provide a constant contact pressure during the winding process, the tension T must be adjusted.

The result of uneven pressure across the wound layer thickness is the binder migration from deeper layers to the outer ones. This leads to the CM creation with different proportions of binder. Under CM part loading with an inhomogeneous structure, the material layers are included in the load support non-simultaneously, that negatively affects its mechanical properties. The binder extrusion from the deep layers is typical for the "wet" winding process. For the "dry" winding process, the solvent is removed from the compound during preliminary tape drying in the heater and the thermosetting binder is partially polymerized. Therefore, in the "dry" winding process, the binder can almost not migrate out from the deep layers to the outer ones.

The main purpose of the tensioning fibers (threads) is their equalization before winding. This makes it possible to simultaneously turn on most of the CM fibers in the load support under mechanical loading. For unidirectional fiberglass, the optimal thread tension is $q_t = (0.05—0.15)q_{br}$, for organo-fibers is $q_t = (0.25—0.36)q_{br}$, for carbon fibers is $q_t = (0.02—0.07)q_{br}$, where q_{br} is the breaking load.

Thread tension is provided in the thread guide (on the way from the spool holder to the part). The tension force is created by all elements of the thread guide. At the same time, special tools are used —tensioners. The reinforcement manufactural tension during the winding process is created by tensioners of various types. Figure 3.17(a) shows roller tensioner. The tape or thread slides over the brake rollers surface. Some force is used to overcome friction. If the position of one of the brake rollers is changed in relation to other, then in this case the contact zone of the reinforcement with the rollers will change ($\varphi_1 > \varphi_2$). The thread tension will change accordingly. Diagrams of other tensioners are shown in Figure 3.17(b) (c). The tension in these devices is adjusted by the brake [Figure 3. 17(b)] or pressure lingoes [Figure 3. 17 (c)]. Mechanical tensioners provide reinforcement tension with an accuracy of ± 0. 2 kg. Brakes adjust the tension on the spools.

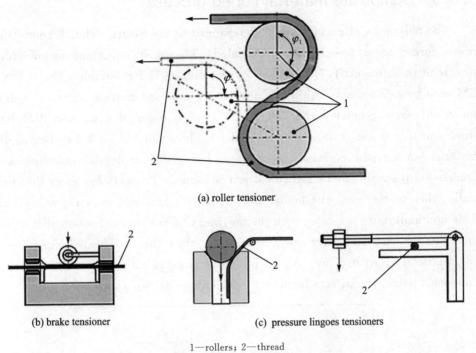

(a) roller tensioner

(b) brake tensioner　　　　　　　　(c) pressure lingoes tensioners

1—rollers; 2—thread

Figure 3. 17　Tensioners of various types

A radius part decrease leads to the tension force increasing. The tension is adjusted with special automated devices.

It is important to provide local tread pressing in winding process time. This problem occurs in cases when the tension force is insufficient to ensure the resulting CM solidity. The process of local pressing is based on the layer molding by using pressure rollers (Figure 3. 18). This approach allows to compact of thick-walled parts. For small diameter parts, a local pressing with one roller is used [Figure

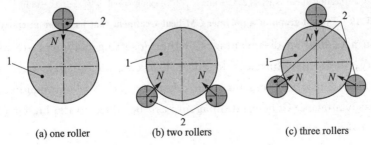

(a) one roller　　　　(b) two rollers　　　　(c) three rollers

1—mandrel; 2—rollers; N—pressing force

Figure 3. 18　Local pressing diagram

3. 18 (a)]. Two- and three-roller pressing can be used with a part diameter increasing [Figure 3. 18 (b) (c)].

3.5.3 Composite material curing process

The polymer CM curing type is determined by its nature. The thermosetting resins curing occurs when they are heated. The most important manufactural feature of these materials is that their curing process is irreversible. The polymer CM curing mode depends on the types of binder and condensation agent, as well as the result parts geometry and weight. A typical graph of a polymer CM heat treatment that is based on an epoxy binder is shown in Figure 3. 19. During the material heating, the curing is accelerated; irreversible molecules crosslinking is occurred in the resin, and a lattice pattern is formed. If a part has great thickness walls, then in this case the binder polymerization proceeds unevenly over time. This non-uniformity is related with the polymer CM low thermal conductivity. The more heated external layers polymerize first, then the less heated inner layers polymerize later with delay. As a non-uniformity polymerization result of large dimension parts, the defects formation (delamination) is possible.

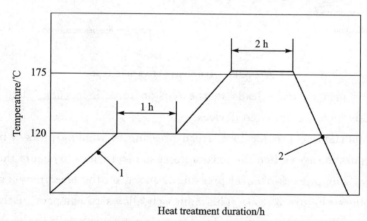

Heat treatment duration/h

1—heating speed is $V_1 = 2$ ℃ · min^{-1}; 2—cooling speed is $V_2 = 2$ ℃ · min^{-1}

Figure 3. 19   Typical graph of a polymer CM heat treatment that based on an epoxy binder

The furnace heating disadvantages of thermosetting polymers are low process productivity, ineffective furnaces loading, low material quality when processing large-dimensional parts. The using of ultrahigh-frequency electromagnetic field has no such disadvantages. Condensation agents can be added to the binder to speed up the polymerization process.

To cure some resins types, it is necessary to provide not only heating, but also pressure on the molded surface. The recommended overpressure value is about

0. 35—0. 7 MPa. The pressure to the part surface is carried out by the molding in an autoclave, the molding with an elastic diaphragm, the molding with compression with a heat-shrinkable tape, the molding under the thermal expansion. The overpressure above the molded part surface provides a CM monolithic structure and increases its physical and mechanical properties.

Thermoplastic polymers curing is carried out when they are cooled (for example, till room temperature). Upon subsequent heating, the thermoplastic material softens again.

3.5.4   Winding pattern types

There are three winding pattern types with according to the kinematic feature: turning, grinding and lapping.

In the turning type, the main movement is the mandrel turning and the feed movement that parallel to the part axis is carried out by head with reinforcement. To implement this process, modernized turning machine can be used [Figure 3. 20 (a)].

In the grinding type, the feed movement provides by the turning mandrel.

In the lapping type [Figure 3. 20 (b)], head with reinforcement must be installed on a swivel, the turning of which is the main movement. The feed movement is carried out by the mandrel longitudinal movement. To tension force stabilizing and improving the winding quality, for this type, guide rollers or pegs are used.

According to the reinforcement fibers packing type, there are several winding ways:

1) hoop;
2) tangential;
3) spiral-cross (spiral-longitudinal, spiral-transverse);
4) combined spiral-ring;
5) longitudinal-transverse;
6) oblique longitudinal-transverse;
7) planar (pole, orbital, planar);
8) tetranocoil;
9) zonal winding.

The choice of the reinforcement lay-up type and its winding approach depends on many factors, including the part operational conditions, its structure, dimensions, the parts number in the batch, etc. Large-sized cylindrical parts are recommended to be made by the hoop winding from pre-impregnated clothes

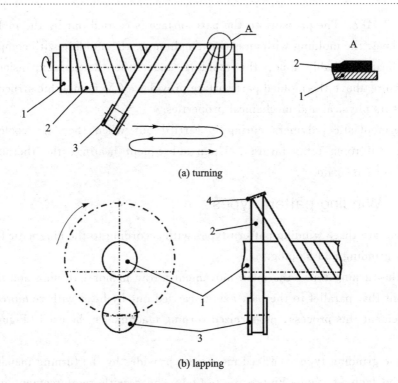

(a) turning

(b) lapping

1—mandrel; 2—reinforcement; 3—head with reinforcement; 4—guide peg

Figure 3. 20   Kinematic winding pattern types

materials. For the large diameter pipes manufacturing, the longitudinal-transverse winding of types from unidirectional fibrous materials can be used.

3.6   Polymer composite material structure production on a base of the pultrusion

Pultrusion is a manufactural process for the polymer CM parts production, which is based on continuous reinforcement pulling that impregnated with a binder through a heated drawhole. The workhole of the drawhole corresponds to the part shape. The pultrusion process diagram is shown in Figure 3. 21(a). The parts cross-sectional shapes produced by pultrusion are shown in Figure 3. 21(b).

Threads, yarns, clothes tapes are used as a reinforcement for the pultrusion. The reinforcement which comes from the spools holder into a tank with a binder, is impregnated with a binder and, through the guide rollers, is pulled into the preforming die, where the substance gradually takes the part shape (in cross-section). From the preforming die, the substance is transferred to the heated mold. In it, the part takes the final required cross-section and the binder polymerized. The final part hardening is done in the oven. At the last stage, the

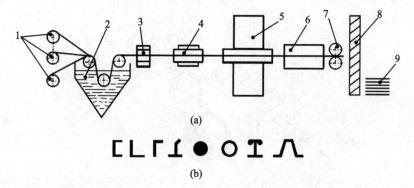

(a)

(b)

1—reinforcement spools; 2—tank with binder; 3—guide rollers; 4—die; 5—heated mold;

6—hardening oven; 7—haul-off unit; 8—cutting tool; 9—store holder

Figure 3. 21   Polymer CM parts production on a base of the pultrusion

part is cut into sections of the required length.

Unlike extrusion, pultrusion does not push the product through the dies, but is pulled out by applying a pulling force. The pulling force is provided by the haul-off unit.

Reinforcement impregnation during pultrusion can be carried out in a tank with a binder located before the preforming die, as it is shown in Figure 3. 21(a). The second way to product parts is to perform the part cross-section with dry fibers and then impregnate it in a drawhole.

An important step in pultrusion is the part hardening. This stage duration determines the pultrusion performance. The part pooling speed is about 0. 6—1. 5 $m \cdot min^{-1}$ that depends on the type of binder and the result part thickness. The curing thermosetting matrices stage is realized in tunnel oven or directly in the drawhole by an ultrahigh frequency electromagnetic field application (Figure 3. 22). The last heating way is more efficient.

The long final part is divided into required lengths sections by cutting tools based on super hard materials (diamonds, carbides, etc.). In some cases, hydraulic cutting is used.

The CM pultrusion with a thermoplastic polymer matrix has significant manufactual differences from the production of polymer CM with a thermosetting matrix. These differences are related with the heat treatment modes and the reinforcement impregnation features. The reinforcement impregnation is carried out directly in the drawhole for the thermoplastic polymers. The part cooling is also in the drawhole, otherwise it will not harden and will lose its shape after ejecting from the drawhole. The part in the drawhole is cooled to the matrix glass-transition region. Thus, the drawhole structure is more complex for the polymer

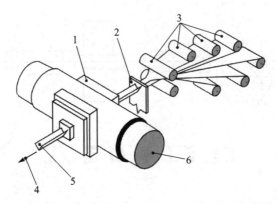

1—furnace with drawhole; 2—preform drawhole; 3—basic substance; 4—pulling force;

5—result hardened part; 6—microwave magnetron

Figure 3. 22   Pultrusion tool diagram with ultrahigh frequency electromagnetic field unit

CM production based on thermoplastic resins than when using thermosetting resins as a matrix material.

The forming pressure in the drawhole is created by decreasing the channel cross-section along the length of the forming zone (impregnation zone), by increasing the substance pulling speed through the drawhole, and by some substance excess. For the polymer CM production with thermoplastic matrices, it is recommended to provide an excess of fibers in the amount of about 5%—15% of the calculated content.

Pultrusion provides the following advantages of polymer CM: high accuracy; the ability to long parts manufacturing; high material utilization; high process productivity; high material structure quality (precise of the ratio between the reinforcement and binder). It becomes possible to reform parts when thermoplastic polymers are used as a matrix material.

3.7 Preform substance processing for polymer composite materials parts production

In order to reduce the manufacturing parts cost from polymer CM, to increase their forming processes productivity, in some cases, special blanks are used. The shape and dimensions of the blanks correspond to the dimensions of future parts. Several approaches for the blanks manufacturing have been developed and used:

1) air pumping from the grid mold;

2) liquid pumping from a perforated mold;

3) spraying;

4) centrifugal filtration;

5) pulp outflow and vacuum dehydration.

The mats obtained by one of the above approaches are further used for the polymer CM parts manufacturing. The mats forming diagrams in molds are shown in Figure 3.23. One of them corresponds to the upper (a), and the other —to the lower (b) mold position. The molding mats scheme with a lower mold position is more often used.

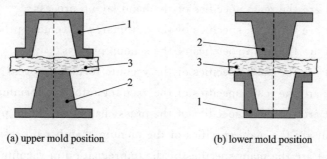

(a) upper mold position   (b) lower mold position

1—mold; 2—punch; 3—mat

Figure 3.23   Mats forming diagrams in molds

Blanks are usually placed on a punch, as shown in Figure 3.24. The binder location depends on the part size. In the case that the part has small dimensions, it is recommended to place the matrix substance on blank top locally. The resin is evenly distributed over the blank surface and impregnates it when the mold will be pressed by the punch.

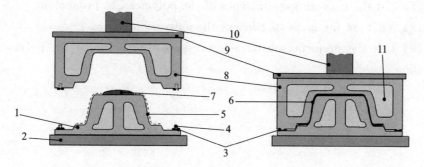

1—punch; 2—bottom support; 3—stop plates; 4—guides; 5—blank; 6—result part;
7—resin; 8—mold; 9—top support; 10—transfer plunger; 11—stream channel

Figure 3.24   CM part forming by mold

If the part is large, then the binder must be placed on the blank not locally, but preliminarily distributed over several zones. In each specific case, the binder location on the blank surface is solved especially by taking into account the result part shape and size.

Questions

(1) List the general operations of the polymer CM production.

(2) List the main forming methods of the polymer CM production.

(3) Briefly describe the features of the impression molding process.

(4) What are the main specifics of the hand lay-up process?

(5) What are the main specifics of the automated placement process?

(6) What are the main specifics of the schoop process?

(7) What are the main specifics of the vacuum forming process?

(8) What are the main specifics of the vacuum-autoclave forming process?

(9) What are the main specifics of the press-chamber molding process?

(10) What are the main specifics of the plenum-chamber process?

(11) What are the main specifics of the impregnation in vacuum process?

(12) What are the main specifics of the direct compression molding process?

(13) What are the main specifics of the injection compression molding process?

(14) What are the main specifics of the thermocompression molding process?

(15) List the winding process types of the polymer CM production.

(16) What are the main specifics of the contact pressure for winding process?

(17) What are the main specifics of the curing process?

(18) List the winding pattern types of the polymer CM production.

(19) What are the main specifics of the pultrusion process?

(20) List the preform substance processing approaches of the polymer CM production.

Chapter 4　Mechanical Behavior of Polymer Composite Materials

4.1　Composite materials modulus of elasticity

The most important features of the CM mechanical behavior are elastic moduli. The elastic moduli are the factors in the relationships that link strains and applied mechanical stresses. The modulus of elasticity, which characterizes the material rigidity under normal stress conditions, is called Young's modulus (normal elastic modulus):

$$E = \frac{\sigma}{\varepsilon}, \tag{4.1}$$

where σ is tensile (compression) stress; ε is specific elongation (compression).

The material rigidity under shear stress is characterized by shear modulus:

$$G = \frac{\tau}{\gamma}, \tag{4.2}$$

where τ is tangential stress; γ is shear strain.

The relationship between the stresses acting on the body and the elastic strains that accrue in it is established by Hooke's equation. If the body is isotropic and only tensile or compressive stress acts on it, Hooke's equation has the view:

$$\sigma = E \cdot \varepsilon. \tag{4.3}$$

In addition to uniaxial tension or compression, such forces application schemes are possible, as a result of which an isotropic body will be in a flat and volumetric stress conditions. In the general case, for an isotropic material, Hooke's equation that establishes the relationship between stresses and strains has the view as follows:

$$
\begin{cases}
\varepsilon_x = \dfrac{1}{E}[\sigma_x - \nu(\sigma_y + \sigma_z)]; \\[2mm]
\varepsilon_y = \dfrac{1}{E}[\sigma_y - \nu(\sigma_x + \sigma_z)]; \\[2mm]
\varepsilon_z = \dfrac{1}{E}[\sigma_z - \nu(\sigma_x + \sigma_y)]; \\[2mm]
\gamma_{xy} = \dfrac{1}{G}\tau_{xy}; \\[2mm]
\gamma_{yz} = \dfrac{1}{G}\tau_{yz}.
\end{cases}
\tag{4.4}
$$

where E is normal elastic moduli in the direction of the axis indicated in the indices; G is shear moduli along the planes indicated in the indices; ν is Poisson ratio in the direction of the first of the axes indicated in the indices, under the normal stresses action in the direction of the second axis.

The simplest thing is to analyze the isotropic materials of elastic behavior. The elastic moduli number of the anisotropic materials behavior reaches 21 (depending on the object symmetry). Most composites are anisotropic. In this regard, the problem of describing their behavior in the strain elastic zone is, as a rule, rather complicated.

Hooke's equation for anisotropic bodies, including composites, has tensor and matrix view. The elastic anisotropic behavior of bodies with a high symmetry degree is described by the Hooke's equation technical view. So, for example, for an orthotropic material, Hooke's equation in technical view has eguation (4.5).

$$
\begin{cases}
\varepsilon_x = \dfrac{\sigma_x}{E_x} - \left(\dfrac{\nu_{yx}}{E_y}\right)\sigma_y - \left(\dfrac{\nu_{zx}}{E_z}\right)\sigma_z ; \\[2mm]
\varepsilon_y = -\left(\dfrac{\nu_{yx}}{E_x}\right)\sigma_x + \left(\dfrac{1}{E_y}\right)\sigma_y - \left(\dfrac{\nu_{zx}}{E_z}\right)\sigma_z ; \\[2mm]
\varepsilon_z = -\left(\dfrac{\nu_{xz}}{E_x}\right)\sigma_x - \left(\dfrac{\nu_{yz}}{E_y}\right)\sigma_y + \left(\dfrac{1}{E_z}\right)\sigma_z ; \\[2mm]
\gamma_{xy} = \dfrac{1}{G_{xy}}\tau_{xy} ; \\[2mm]
\gamma_{yz} = \dfrac{1}{G_{yz}}\tau_{yz} ; \\[2mm]
\gamma_{xz} = \dfrac{1}{G_{xz}}\tau_{xz} ,
\end{cases}
\tag{4.5}
$$

In this set of equations, there are 12 elastic moduli. Nine of them are independent. If anisotropic materials with a higher symmetry degree are analyzed, the independent elastic constants number decreases.

One of the simplest CM types is composites with continuous fibers oriented in one direction. The elastic feature analysis of such materials is a relatively simple task. Traditional approaches to CM as structurally inhomogeneous materials make it possible to calculate their elastic moduli from known elastic properties and volumetric components concentrations.

Here, for the elastic behavior analyzing of unidirectional CM, it will be considered two extreme options cases related with material loading in the direction of the fiber axes and in the transverse direction.

4.1.1 Normal elastic modulus of one-directional composite material in the direction of the reinforcement axis

As the simplest object for analysis, a monolayer is usually taken, which is reinforcement fibers laid up in one direction, which are bended by a matrix material. The example in Figure 4. 1 has, for analysis ease, the elements of the fiber and matrix with rectangular shape cross-section.

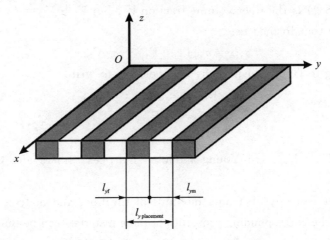

l_{yf}—the width of fiber; l_{ym}—the width of matrix layer; $l_{y\,placement}$—the width of fiber and matrix placement.

Figure 4.1 Diagram of unidirectional fibers CM

If the analyzed plate is loaded in the longitudinal direction, that is, along the x axis, by the force P_x, then the compatibility of the fibers and matrix elements strains makes it possible to write

$$\varepsilon_{xcm} = \varepsilon_{xm} = \varepsilon_{xf}, \tag{4.6}$$

where ε_{xcm}, ε_{xm}, ε_{xf} are the specific strain values of the CM, matrix and fiber in the direction of the x axis.

The total force P_x that acting on the plate can be represented as the sum of the force P_{xm} that acting on the matrix and the force P_{xf} that acting on the fiber

$$P_x = P_{xm} + P_{xf}. \tag{4.7}$$

The force P, which included in the dependence (4.7), can be expressed from the equations that connected them with mechanical stress σ,

$$\sigma = \frac{P}{S}, \tag{4.8}$$

where S is cross-section area of the element that tensed by force P.

Taking into account equation (4.8), dependence (4.7) can be written in the following view:

$$\sigma_{xcm} \cdot S_{cm} = \sigma_{xm} \cdot S_m + \sigma_{xf} \cdot S_f, \tag{4.9}$$

where σ_{xcm}, σ_{xm}, σ_{xf} are tensile stresses in the CM, matrix and fiber, respectively, in the direction of the x axis; S_{cm}, S_m, S_f are cross-section areas of the CM, matrix and fiber, respectively.

We divide both sides of equation (4.9) by the value of S_{cm}:

$$\sigma_{xcm} = \sigma_{xm} \frac{S_m}{S_{cm}} + \sigma_{xf} \frac{S_f}{S_{cm}}. \tag{4.10}$$

Considering that, (S_m/S_{cm}) is the matrix volume fraction V_m in the analyzed CM and (S_f/S_{cm}) is the fiber volume fraction V_f, but $V_f + V_m = 1$, then equation (4.10) can be transformed to:

$$\sigma_{xcm} = \sigma_{xm}(1 - V_f) + \sigma_{xf}V_f. \tag{4.11}$$

According to Hooke's equation, it is possible to write

$$\begin{cases} \sigma_{xm} = \varepsilon_{xm} E_{xm}; \\ \sigma_{xf} = \varepsilon_{xf} E_{xf}; \\ \sigma_{xcm} = \varepsilon_{xcm} E_{xcm}. \end{cases} \tag{4.12}$$

where E_{xm}, E_{xf}, E_{xcm} are Young modulus matrices, fibers and CM in the x direction.

Due to the isotropy of Young modulus of the fiber and matrix, the direction index x, y of the corresponding properties of these materials can be omitted. Taking into account dependences (4.6) and (4.12), equation (4.11) takes the view:

$$E_{xcm} = E_f V_f + E_m V_m = E_f V_f + E_m (1 - V_f). \tag{4.13}$$

Thus, knowing the volumetric content of the matrix and fiber in the CM, as well as the elastic moduli of these components, in accordance with equation (4.13), it is possible to calculate the elastic modulus of the unidirectional CM in the direction of the reinforcement axis.

4.1.2 Normal elastic modulus of one-directional composite material in the transverse direction of the reinforcement axis

Let the CM in the conditions of a monolayer (see Figure 4.1) be loaded with a force P_y applied transverse to the fibers longitudinal axis. The normal stress σ, that occurs in the analyzed object, is the ratio of the force P_y to the area of the object S in the cross-section transverse to the y-axis. Due to the equality of the forces acting on the fiber, the matrix and the CM as a whole and the equality of the areas on which these forces act, it is possible to write

$$\sigma_{yf} = \sigma_{ym} = \sigma_{ycm}. \tag{4.14}$$

The absolute CM strain in the direction of the y-axis Δ_{ycm} will be equal to the sum of the absolute strains of the fibers and matrix:

$$\Delta_{ycm} = \Delta_{yf} + \Delta_{ym}. \tag{4.15}$$

Substitute in equation (4.15) the corresponding values $\Delta = l\varepsilon$, where l is the length, and ε is the specific strain of the deformed element. As a result, there is the dependence:

$$\varepsilon_{ycm} l_{ycm} = \varepsilon_{yf} l_{yf} + \varepsilon_{ym} l_{ym}, \tag{4.16}$$

where ε_{ycm}, ε_{yf}, ε_{ym} are specific strains respectively of the CM, fibers and matrix in the direction of the y-axis; l_{ycm}, l_{yf}, l_{ym} are length respectively of the analyzed CM element, fibers and matrix in the direction of the y-axis.

We divide both sides of equation (4.16) by the l_{ycm} and as a result, we obtain a dependence of the view

$$\varepsilon_{ycm} = \varepsilon_{yf} \frac{l_{yf}}{l_{ycm}} + \varepsilon_{xm} \frac{l_{ym}}{l_{ycm}}. \tag{4.17}$$

Considering that, (l_{yf}/l_{ycm}) and (l_{ym}/l_{ycm}) are actually fiber and matrix volume fractions V_f and V_m in the analyzed monolayer, but $V_f + V_m = 1$, then equation (4.17) can be transformed to

$$\varepsilon_{ycm} = \varepsilon_{yf} V_f + \varepsilon_{ym} V_m = \varepsilon_{yf} V_f + \varepsilon_{ym} (1 - V_f). \tag{4.18}$$

In accordance with Hooke's equation, the strain of an object is determined by

$$\varepsilon = \frac{\sigma}{E}. \tag{4.19}$$

Substitute into equation (4.18) the corresponding values of the specific strain, we obtain

$$\frac{\sigma_{ycm}}{E_{ycm}} = \frac{\sigma_{yf}}{E_{yf}} V_f + \frac{\sigma_{ym}}{E_{ym}} (1 - V_f). \tag{4.20}$$

Taking into account dependence (4.14), which makes it possible to reduce all stress values in both parts, the equation that describing the normal elastic modulus of a unidirectional CM in the direction transverse to the reinforcement axis can be written as

$$E_{ycm} = \frac{E_f E_m}{E_m V_f + E_f (1 - V_f)}. \tag{4.21}$$

The indices$y$ for the elastic moduli values of the fiber and matrix are omitted in the last equation, since it is assumed that the materials of both components are isotropic.

Equations (4.13) and (4.21) are very convenient for describing the elastic properties of CM, but at the same time they are considered only estimates, approximate, since they are based on the use of idealized models.

4.2　Composite materials tensile strength

4.2.1　Breaking strength of an unidirectionally reinforced composite material

One of the simplest cases for analyzing the CM strength properties can be the tensile of a unidirectional CM with reinforcement of continuous fibers. The tensile force is directed along the fiber axes.

The analysis of the behavior under tensile force conditions of the CM as a whole, as well as separately of the fibers and matrix, makes it possible to schematically represent the tensile diagrams of these materials in one figure (Figure 4.2).

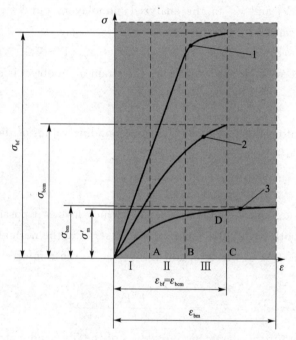

1—fibers; 2—unidirectional reinforced CM; 3—matrix

Figure 4.2　Tensile diagram

Let the CM consists of a soft plastic matrix and harder and stronger fibers. The material elastic modulus in the tensile diagram is characterized by the slope ratio to the strain axis. The tensile diagrams of such CM and its components are shown in Figure 4.2. The CM behavior is explained by curve 2, the slope of which is in the interval between the slopes of the tensile curves of fiber 1 and matrix 3.

In the general case, three sections (I , II , III) are intercepted on the CM tensile diagram. Section I corresponds to the elastic behavior of both the fiber and the matrix. In section II , the fiber continues to deform elastically, and the matrix behaves elastically-plastically. Within section III , both the fiber and the matrix are plastically deformed. The breaking strain of the analyzed CM actually corresponds to the fiber breaking strain, i. e. , the material breaks at the moment when the spontaneous reinforcement destruction starts. In this case, the matrix plastic properties, judging by Figure 4. 2 are not exhausted.

If we use the elastic modulus analysis of a unidirectional CM loaded along the fiber axis and take equation (4. 11), then the analyzed CM tensile strength can be written in the following view:

$$\sigma_{bcm} = \sigma_{bf} V_f + \sigma'_m (1 - V_f), \tag{4.22}$$

where σ_{bf} is fibers tensile breaking stress; σ'_m is matrix tensile stress in time fibers destruction in the CM.

The value of σ'_m can be determined like in Figure 4. 2. For the plastic low strengthen matrices the value of σ'_m can be taking like the matrix breaking stress σ_{bm}.

Dependence (4. 22) is well-known as the rule of mixtures. Equation (4. 22) based on the assumption that the bond strength at the connection between the fiber and the matrix is sufficient to ensure joint deformation of these components up to complete CM destruction. In addition, it was assumed that the CM destruction occurs, with the simultaneous, all its fibers destruction. In real CM, these assumptions may not be met. These deviations from the idealized model of CM destruction explain the cases of non-observance to one degree or another of the rule of mixtures observed in practice. Nevertheless, it is believed that dependence (4. 22) is quite suitable for performing estimated calculations of the CM strength properties.

4.2.2　Fiber orientation influence on composite material strength properties

The vast majority of CM is anisotropic materials, the properties of which largely depend on the external forces action direction. Here, as a relatively simple example, we consider that the unidirectional fibrous material behavior depends on the fibers orientation in relation to the tensile force (Figure 4. 3). It is known that, three CM fracture modes are possible depending on the angle θ between the fiber axis and the tensile force direction.

At low values of θ that are not exceeding a few degrees, the CM is destroyed as a result of the fibers breaks under the normal stresses action. If the matrix strength can be neglected, the CM ultimate strength is described by the following equation:

$$\sigma_\theta = \frac{\sigma_{bf}}{\cos^2\theta}, \qquad (4.23)$$

where σ_{bf} is ultimate fibers strength.

Graphical dependence (4.23) is shown in Figure 4.4. With an angle θ increasing, the fibers tensile strength increases. It was found experimentally that dependence (4.23) is fulfilled only at low angle θ.

At high angle θ, close to 90°, the CM strength is determined by the normal strength of the boundary between the matrix and the fibers or the matrix strength under conditions of its tension. In the latter case, the CM ultimate strength is described by the dependence

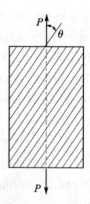

Figure 4.3　Load configuration of a unidirectional CM at an angle θ to the fiber axis

$$\sigma_f = \frac{\sigma_{bm}}{\sin^2\theta}, \qquad (4.24)$$

where σ_{bm} is ultimate matrix strength in plain strain conditions.

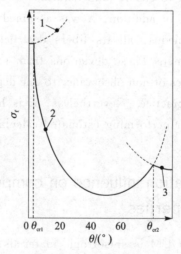

1—section with $\sigma_f = \dfrac{\sigma_{bf}}{\cos^2\theta}$; 2—section with $\sigma_f = \dfrac{\tau_m}{\sin\theta\cos\theta}$; 3—section with $\sigma_f = \dfrac{\sigma_{bm}}{\sin^2\theta}$

Figure 4.4　Dependence of the unidirectional CM ultimate strength on fiber orientation

In the interval between the critical angles θ_{cr1} and θ_{cr2}, that correspond to the dependences (4.23) and (4.24), the CM strength is described by the dependence

$$\sigma_f = \frac{\tau_m}{\sin\theta\cos\theta} \qquad (4.25)$$

where τ_m is ultimate matrix or boundary (between matrix and fibers) strength in shear conditions.

Thus, at $\theta_{cr1} < \theta < \theta_{cr2}$, the CM strength depends on the fracture mode that is based on the shear stresses and the matrix destruction or the boundary between the matrix and the fibers as a result of shear along planes parallel to the fibers.

Equating dependences (4.23) and (4.25), it is possible to find the critical angle θ_{cr1},

$$\tan \theta_{cr1} = \frac{\tau_m}{\sigma_{bf}}. \tag{4.26}$$

Similarly, one can find the θ_{cr2}, and for this it is necessary to equate dependences (4.24) and (4.25)

$$\tan \theta_{cr2} = \frac{\sigma_{bm}}{\tau_m}. \tag{4.27}$$

4.2.3 Composite material strength properties reinforced with discrete fibers

This section deals with CM with discrete (short) fibers oriented in one direction. A tensile force is applied along the fiber axes. It is necessary to apply the critical fiber length concept for the behavior analyzing of such fibers under tensile stresses.

In a CM, a tensile load is transferred to the fibers through the matrix due to shear stresses acting at the boundary between its components. The fibers tips are not loaded. The fibers central sections are most heavily loaded. If a fiber degenerates into a small spherical particle, then it can be conventionally considered that it does not carry a tensile load. In alloys with dispersed spherical particles under the tensile stresses action, the particles destruction is usually not present. Particles peel away from the matrix is typical.

It also happens on CM. Under the tensile load action, discrete fibers are not destroyed, but pulled out of the matrix. Fibers pulling out without their destroying is possible only if their length l is shorter than the critical length l_{cr}. At $l > l_{cr}$ under the tensile force action, the fibers are destroyed. Thus, the fiber critical length l_{cr} is the length at which it breaks down in a CM.

Given the assumption that the shear stresses are uniformly distributed along the fibers length, the critical fiber length can be quantified. This calculation is based on the equality of the tangential and normal forces that acting on the fiber,

$$\sigma_f \frac{\pi d_f^2}{4} = \tau \pi d_f \frac{l}{2} \tag{4.28}$$

where σ_f is normal tensile stress in the fiber; d_f is fiber diameter; τ is shear stress at the boundary between the matrix and the fibers; l is fiber length.

The fiber critical length can be calculated by

$$l_{cr} = \frac{\sigma_{bf}}{2\tau_{bound}} d_f,$$

$$(4.29)$$

where τ_{bound} is boundary strength between the matrix and the fibers; σ_{bf} is ultimate normal tensile stress in the fiber.

Analysis of dependence (4.29) shows that the critical fiber length increases with an increase in the fiber ultimate strength, a decrease in the shear boundary strength, and an increase in the fiber diameter.

The CM strength that reinforced with discrete fibers depends on the length of the fibers. For $l < l_{cr}$, the rule of mixtures equation has the view

$$\sigma_{bcm} = \frac{\tau_{bound} l}{d_f} V_f + \sigma_{bm}(1 - V_f).$$

$$(4.30)$$

The CM ultimate strength containing discrete fibers with a length exceeding the critical one ($l > l_{cr}$) is described by the following dependence

$$\sigma_{bcm} = \sigma_{bf}\left[1 - (1 - k)\frac{l_{cr}}{l}\right] V_f + \sigma'_m(1 - V_f),$$

$$(4.31)$$

where k is factor, $k < 1$.

With an increase in the ratio l/l_{cr}, the CM strength increases. In the event that the fiber length is 10 times longer than the critical one, the CM ultimate strength with discrete fibers is only 5% less than the CM ultimate strength reinforced with continuous fibers.

4.2.4　Fiber volume fraction influence on composite material strength

The fiber volume fraction has a significant effect on the CM strength. In accordance with the rule of mixtures, the stronger fiber volume fraction is, the higher CM ultimate strength is. Reasoning from this fact, one should maximally increase the fiber volume fraction in the material. However, in practice, the maximum fiber fraction, in order to avoid delamination of the CM does not exceed 0.7.

In addition, it should be said that there are also geometric restrictions on the maximum fiber volume fraction in the CM. So, for example, using fibers of circular cross-section, the tetragonal fibers lay-up provides the maximum of their volume fraction $V_{max} = 0.785$ (Figure 4.5).

For the case of a denser fibers hexagonal lay-up $V_{max} = 0.907$ [Figure 4.5

(b)], and when using fibers of two diameters $V_{max} = 0.924$ [Figure 4.5 (c)]. These fiber volume fractions correspond to the fibers lay-up without the clearance formation between them.

It is necessary to provide that a matrix interlayer for CM delamination risk reduces between the fibers. If the interlayer matrix of minimum thickness δ_{min} presences, the fiber volume fraction in the CM with tetragonal lay-up [Figure 4.5 (c)] is

$$V_f = \frac{\pi}{4 + 4(\delta_{min}/d_f) + (\delta_{min}/d_f)^2},\tag{4.32}$$

with hexagonal lay-up [Figure 4.5 (e)]

$$V_f = \frac{2\pi d_f^2}{(2d_f + \delta_{min})^2 \sqrt{3}}.\tag{4.33}$$

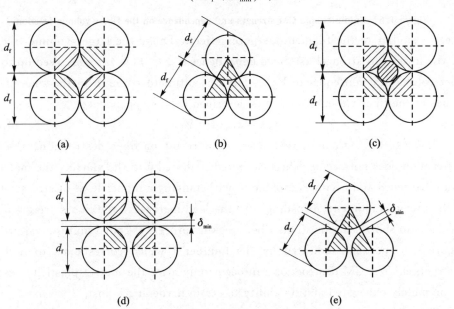

Figure 4.5   Unidirectional fibers lay-up configurations in direct contact [(a) (b) (c)] and lay-up with matrix interlayers [(d) (e)]

Thus, on one hand, in real CM, there are limitations on the maximum fiber volume fraction. On the other hand, in CM with a plastic matrix and brittle fibers, the minimum fiber volume fraction is also limited. The meaning of these restrictions is clear from Figure 4.6, which shows the CM strength and dependence on the fibers volume fraction.

It can be seen from the given figure that there is a value range of the fiber volume fraction, in which the CM strength σ_{bcm} decreases with increasing V_f (Figure 4.6, zone I). This CM behavior is explained by the fact that at low V_f, the fibers

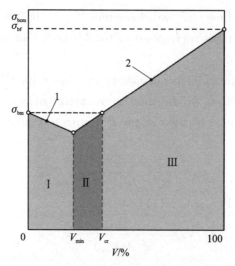

Figure 4.6   Unidirectional CM strength and dependence on the fibers volume fraction

effect is similar to the filamentous pores effect. Thus, the larger volume fraction is, the lower CM strength is. Such dependence $\sigma_{bcm} = f(V_f)$ is preserved up to the fiber volume fraction equal to V_{min}. The CM ultimate strength in the zone of low values of the fiber volume fraction is described by the dependence

$$\sigma_{bcm} = \sigma_{bm}(1 - V_f). \tag{4.34}$$

If $V_f < V_{min}$, then in this case, as a result of fiber destruction, the CM destruction does not occur. The load is redistributed into the matrix; the matrix is strain hardened and becomes capable of withstanding a higher level of stresses than at the previous moment of loading. As the load increases, the fibers are gradually divided into short sections. This fiber cracking process is called multiple failure. In Figure 4.6 the zone indicated by the number I which corresponds to multiple destruction. The CM destruction can occur only after the matrix plasticity reserve is completely exhausted and its ability to strain hardening is lost.

The minimum fiber volume fraction V_{min} is the volume fraction that corresponds to the lowest CM ultimate strength level. At $V_f > V_{min}$, an ascending section of the dependence $\sigma_{bcm} = f(V_f)$ is observed

$$\sigma_{bcm} = \sigma_{bf}V_f + \sigma'_m(1 - V_f). \tag{4.35}$$

It is necessary to equate the right-hand sides of equations (4.34) and (4.35) solve the resulting equation for the value V_{min} to quantitatively determine the minimum fiber volume fraction

$$\sigma_{bm}(1 - V_{fmin}) = \sigma_{bf}V_{fmin} + \sigma'_m(1 - V_{fmin}). \tag{4.36}$$

$$V_{fmin} = \frac{\sigma_{bm} - \sigma'_m}{\sigma_{bf} + \sigma_{bm} - \sigma'_m}. \tag{4.37}$$

From Figure 4.6 it follows that from a practical point of view, it makes no sense to develop and use CM with a minimum fiber volume fraction, since the strength of such materials is lower than the strength of even a plastic matrix, i. e. $\sigma_{bcm} < \sigma_{bm}$. The second characteristic value for the dependence $\sigma_{bcm} = f(V_f)$ is the critical fiber volume fraction V_{cr}. This value corresponds to such a fiber volume fraction at which the CM strength becomes equal to the matrix strength. Naturally, in practice it makes sense to use such composites, for which $V_f > V_{cr}$. It is necessary to equate the right side of equation (4.35) with the unreinforced matrix ultimate strength to quantify the value of V_{cr}

$$\sigma_{bf} V_{cr} + \sigma'_m (1 - V_{cr}) = \sigma_{bm}. \tag{4.38}$$

The equation solution (4.38) for V_{cr} gives

$$V_{cr} = \frac{\sigma_{bm} - \sigma'_m}{\sigma_{bf} - \sigma'_m}. \tag{4.39}$$

In the case that the CM contains a brittle matrix and plastic fibers, the dependence $\sigma_{bcm} = f(V_f)$ will have the other form (Figure 4.7). This relationship consists of two ascending zones. It makes no sense to talk about the values of V_{min} and V_{cr} in this case.

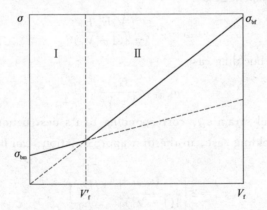

Figure 4.7   CM strength dependence with a brittle matrix and plastic
fibers on the fibers volume fraction

4.3   Composite materials compression strength

Continuous fiber-reinforced CM generally should not operate under compressive stress conditions. Nevertheless, it makes sense to consider the CM behavior under these loading conditions. If the CM is presented in the form of elastic fibers (plates) with unlimited strength that located in a plastic matrix, then the material destruction occurs because of the fibers buckling.

It is possible to use two destruction modes of the CM. In the case that the fibers are bent in opposite directions and a symmetrical case of their buckling is present, the material destruction is the result of matrix stretching in the direction perpendicular to the fiber axis [Figure 4.8 (a)]. If the fibers are bent in one direction and there is an asymmetric buckling case [Figure. 4.8 (b)], the CM destruction is the shear matrix deformation.

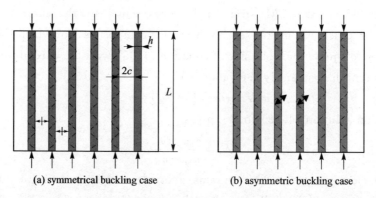

(a) symmetrical buckling case　　　　　　(b) asymmetric buckling case

Figure 4.8　2D diagram of a heterophase layered material compression

The compressive stress that corresponds to the fiber symmetric buckling is determined by the dependence

$$\sigma_{\text{buckl}} = 2V_f \sqrt{\frac{V_f E_m E_f}{3(1-V_f)}}, \tag{4.40}$$

for the asymmetric buckling case

$$\sigma_{\text{buckl}} = \frac{G_m}{1-V_f}. \tag{4.41}$$

The CM critical strain ε_{cr}, corresponding to its destruction according to the symmetric fiber buckling case, in the first approximation, can be calculated by the equation

$$\varepsilon_{\text{cr}} = 2 \left[\frac{V_f}{3(1-V_f)} \right]^{0.5} \left(\frac{E_m}{E_f} \right)^{0.5}, \tag{4.42}$$

for the asymmetric buckling case

$$\varepsilon_{\text{cr}} = \left[\frac{1}{V_f(1-V_f)} \right] \frac{G_m}{E_f}. \tag{4.43}$$

It is necessary to increase the fibers and the matrix rigidity, to increase strength of the boundary between the fiber and the matrix for the CM strength increasing in compression.

4.4　Composite materials destruction features

The most important feature of the structural materials behavior is their crack resistance (fracture toughness), that is the ability to resist crack creation. There are always cracks, pores or other discontinuities kinds in real structures. There are questions: what are sizes of such discontinuities and what is the resistance to these defects? Failure mechanics investigates the behavior of materials containing cracks.

The CM important advantage containing high-strength reinforcement and a viscous matrix is the effective stresses redistribution during the separate fibers destruction. This ensures a high level of CM fracture toughness.

The CM destruction, which are clear anisotropic parts, has some features. One of them is that the CM fracture toughness is mostly determined by the crack propagation direction. In the anisotropic CM, there are directions that provide a high and low level of crack resistance. As a rule, the crack develops easily along the boundary between the fibers and the matrix, i. e. , CM are prone to delamination. If the crack develops across the fibers, then it is possible to achieve high level of the material crack resistance. In this regard, the reinforcement lay-up must clearly correspond to the stress conditions that occurs during the part operation.

The most important CM manufactural defects include delamination that occurs at the stages of manufacturing, transportation and operation. This types of defects can be caused by thermal stresses, impacts, and other types of local loading. Surface delamination is bulging of some material layers. Some examples of delamination in CM are shown in Figure 4. 9.

These types of defects can be caused, for example, by compression stresses or surface heating.

CM delamination along the fiber-matrix conjugation boundaries is one of the effective ways for cracks inhibition that develops perpendicular to the fibers. The stress conditions analysis at the crack tip that propagating through the material, carried out by J. Gordon and J. Cook, showed that tensile σ_{zz} and transverse σ_{xx} stresses can be developed at the crack tip (Figure 4. 10). The relationship between these stresses is described by the dependence $\sigma_{zz} = 4\sigma_{xx}$. In the case, that material delamination occurs when the crack comes the boundary, the crack direction changes by 90°, i. e. , the crack is effectively inhibited. The load-carrying capability of the material is kept.

The crack pass propagation during interlayer fracture is determined by the boundary shape between the matrix and the reinforcement. It is believed that this fracture type is the closest in nature to the fracture types that analyzed in fracture mechanics. Therefore, the interlayer fracture processes are described by traditional fracture mechanics approaches that developed for metal structural materials testing. At the same time, the real CM test data show that the results of such calculations should be considered correct only in the preliminary approximation.

CM have a variety of fracture ways. For example, the unidirectional composite

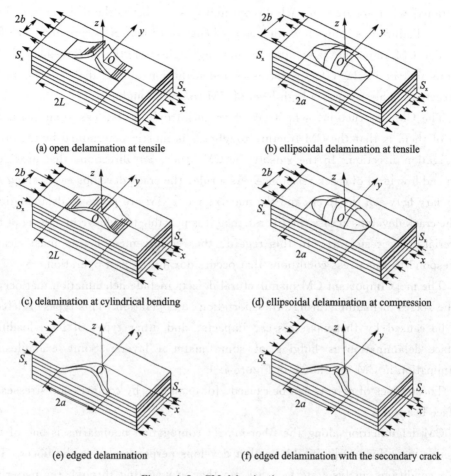

(a) open delamination at tensile   (b) ellipsoidal delamination at tensile

(c) delamination at cylindrical bending   (d) ellipsoidal delamination at compression

(e) edged delamination   (f) edged delamination with the secondary crack

Figure 4.9 CM delamination types

strain may have a result in fiber breakage, cracking at the fibers and the matrix boundary, and matrix destruction. In addition, destruction processes are implemented which combined the listed cases. The two most important disperse damage types for unidirectional CM under tensile stress conditions, namely, single fiber breaks and the matrix and the fiber boundary destruction, are schematically shown in Figure 4.11. The first damage type depends on the ratio of the breaks number to the total structural elements number in

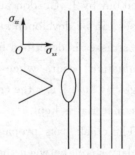

Figure 4.10 The fiber-matrix boundary delamination when a crack interacts with the boundary

the considered CM volume. The second defect type relates with the fiber, and the matrix boundary destruction is defined as the ratio of the sum of damaged boundary sections' lengths to the total fibers' length in the analyzed material volume.

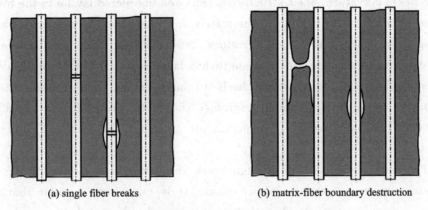

(a) single fiber breaks   (b) matrix-fiber boundary destruction

Figure 4. 11   Unidirectional CM disperse damage types

The CM fracture shape depends on its structure and external loading conditions. In the unidirectional CM tensile case, the fracture surface has brush-like type [Figure 4. 12 (a)]. For the force directed perpendicular to the fiber axis loading case, for example, the three-point bending CM case, fracture has longitudinal cracking [Figure 4. 12 (b)]. In the interlayer delamination case that is shown in Figure 4. 12 (c), the CM fracture is related with the crack development along the fibers and the matrix boundary.

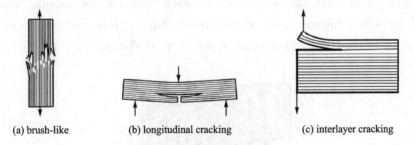

(a) brush-like   (b) longitudinal cracking   (c) interlayer cracking

Figure 4. 12   Laminated CM fracture types

It is usually assumed that the fibers have the given geometric shape and ordered lay-up for the CM behavior analysis under various external loading conditions. In addition, other assumptions are applied to simplify mathematical calculations. For example, it is assumed that the total fibers tensile strength is the same and that their destruction occurs simultaneously when a certain stress level is reached. This approach to the CM fracture process analysis is too simplified. The models which are most suitable for describing such strength features that the rule of mixtures is used, are CM parts consisting of a plastic matrix and strong plastic fibers. Such materials parts are destroyed due to the plastic flow instability with the neck formation with simultaneous total fibers breaking.

In real conditions, due to the matrix and separate fibers properties scatter, due to the fibers curvature, their structure defects and unordered lay-up in the matrix, due to the matrix defects, due to the matrix and fibers boundary defects, the CM mechanical properties have random values. As for brittle fibers located in a low-plastic matrix, as a rule, their strength has large scatter. The strength average values application in calculations leads to huge errors. The refore, the CM structures load-carrying capability forecast should be probabilistic. Considering this, CM stochastic models destruction are used for the loaded CM behavior analysis.

To create materials with a high crack resistance level is one of the most important modern materials science tasks. It is taken into account that crack resistance is an integral value that simultaneously depended on the material strength and plasticity. The higher material strength and plasticity level are, the higher its fracture toughness is. Proceeding from this, the crack resistance increasing task of metal structural materials is usually solved.

CM are differed from metallic materials because they allow applying of additional effective ways for the crack resistance increasing. These ways are related with the boundary between the matrix and the fibers. The boundary is an effective barrier to propagating cracks.

There are at least two ways for the crack resistance increasing, which are typical for CM. Figure 4. 13 shows these ways. Here, schematically crack propagation in the CM perpendicular to the fibers is shown.

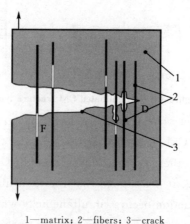

1—matrix; 2—fibers; 3—crack

Figure 4. 13   The crack propagation in a CM, perpendicular to the fibers

One of the crack resistance increasing way is related with the matrix and the fiber boundary destruction. This process requires a certain energy, therefore, leads to the fracture toughness increasing. The boundary destruction process develops near the crack tip. In Figure 4. 13 this zone is marked with D.

The second way for CM fracture toughness increasing is due to the fibers pulling from the matrix. In Figure 4. 13, the fiber pulling zone, located at some distance from the crack tip, is marked with F. The additional work is spent on overcoming the friction forces between the fiber and the matrix, which also results in the CM fracture toughness increasing. This way is effective if the fibers' length is less than the critical length, or equal to it ($l_f \leqslant l_{cr}$).

Otherwise, there is no fibers pulling, but mainly their breaking in the matrix. Graphically, the fracture work W_{fp} depending on the fiber length is shown in Figure 4. 14. This way for the crack resistance increasing is also effective even if both of the CM components (the matrix and the fiber) are brittle. Brittle matrix reinforcement with discrete brittle fibers with a length less than the critical one is a radical way of the CM fracture toughness increasing. In this case, the energy consumption for the fiber pulling from the matrix significantly exceeds the fracture work of each component separately.

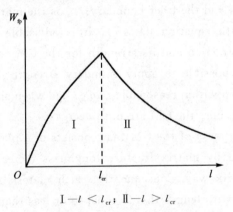

$$I-l < l_{cr}; \quad II-l > l_{cr}$$

Figure 4. 14 The fracture work W_{fp} depending on the fiber length for fibers pulling from the matrix

The energy consumption with the highest level of $W_{fp\ max}$ related with fibers pulling from the matrix occurs if the separate fibers length is equal to the critical length ($l = l_{cr}$):

$$W_{fp\ max} = \frac{1}{12} V_f \sigma_{bf} l_{cr}. \tag{4.44}$$

Analysis of equation (4. 44) shows that in order to increase the fracture work related with the fibers pulling from the matrix, the fibers volume fraction V_f, fibers ultimate strength σ_{bf} and the critical length l_{cr} should be increased. The critical fiber length is proportional to its diameter d_f and inversely proportional to the fiber and the matrix boundary strength τ_{bound} (see equation (4. 29)). So, the CM destruction energy consumption can be increased by fiber diameter increasing and decreasing τ_{bound}. However, in the latter case, there is a CM strength loss under conditions of shear, compression and tension in the direction perpendicular to the fiber axis. Therefore, it is preferable to increase the reinforcement diameter for brittle matrix-brittle fiber fracture toughness increasing. The using of fibers with a diameter of 0. 05—0. 5 mm is considered to manufactural conditions. In cases where separate fibers have optimal parameters, the CM with a ceramic matrix impact strength can be increased tenfold due to the fiber pulling effect.

In the case when the CM consists of a brittle matrix and a plastic fiber, the destruction type largely depends on the fibers effect. A crack, easily developed along a brittle matrix, can be brake by a plastic fiber. The following dependence can be used to calculate the critical stress intensity factor:

$$K_c = \left[\frac{2E_{cm}}{(1-\nu_{cm})^2}(1-V_f)\gamma_m + \frac{V_f\sigma_f^3 d_f}{12\tau E_f} \right]^{0.5}, \tag{4.45}$$

where E_{cm} and ν_{cm} respectively are Young modulus and Poisson ratio for the CM; E_f and σ_f respectively are Young modulus and ultimate stress of fibers; τ is shear strength of the matrix and the fiber boundary; γ_m is the matrix surface energy.

In accordance with equation (4. 45), it is advisable to increase the fiber diameter, its volume fraction and its strength for the CM crack growth resistance increasing. It is also possible to apply boundary strength decreasing. However, this way should be used within reasonable range, and when simultaneously with the crack resistance increasing, the CM strength decreases.

The fracture toughness of the CM that consists of a plastic matrix and brittle fiber can be increased by matrix fracture toughness increasing and using fibers pulling from the matrix way. Consequently, such materials should be consisted from separated fibers with length that is equal to or less than the critical length.

Thus, the CM behavior analysis under external loading indicates that their destruction ways differ from traditional metal structural materials. It is possible to find effective ways to the crack resistance increasing by the using of the production and the CM real structure specifics.

Questions

(1) What are the main mechanical behavior indexes of the composite materials?

(2) Explain equation for the Young modulus (normal modulus of elasticity) E.

(3) Explain equation for the shear modulus G.

(4) Explain relation between stresses and strains on the base of Hooke's equation.

(5) Explain relation between stresses and strains on the base of Hooke's equation for an orthotropic material.

(6) Briefly describe normal elastic modulus of one-directional composite material in the direction of the reinforcement axis.

(7) Briefly describe normal elastic modulus of one-directional composite material in the transverse direction of the reinforcement axis.

(8) Explain breaking strength of an unidirectional reinforced composite material.

(9) Briefly describe fiber orientation influence on composite material strength properties.

(10) Briefly describe composite material strength properties reinforced with discrete fibers.

(11) Briefly describe fiber volume fraction influence on composite material strength.

(12) Briefly describe composite materials compression strength specifics.

(13) Explain composite materials destruction features.

Chapter 5　Composite Materials Joints

According to the mechanical engineering classifications, the CM parts of joints can be classified as detachable and dead, movable and fixed. At the same time, the CM specific performance makes it possible to create structure that can satisfy contradictory and even seemingly mutually exclusive requirements. The requirements severity for CM structures is the main reason for the new technologies development for their manufacture. Therefore, in this chapter, the CM joints part structures are considered in close connection with their purpose and manufactural specifics.

5.1　Composite materials joints classification

CM part joint structures, according to the loads transfer nature, are divided into butt and support. Combined and metal-plastic are most common and are butt joints.

In butt joints, loads of the joint are supported by metal parts and transmitted through the connecting part, which is made from CM.

In the support joints, the load at the joint is transferred directly by the adhesive layers (for example, a skin with a honeycomb core joint). CM parts joints are divided into three classes (Figure 5.1).

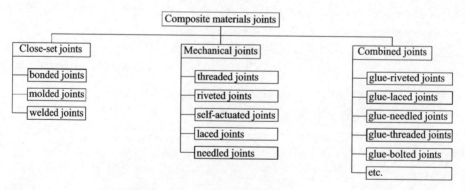

Figure 5.1　CM parts joint classification

They are:

1) Close-set: bonded, molded and welded;

2) Mechanical: threaded, riveted, self-actuated, laced and needled joints;

3) Combined: glue-riveted, glue-laced, glue-needled, glue-threaded, etc.

Among these joints, only mechanical ones are detachable or partially detachable. Partially detachable joints are used not to ensure the disassembly possibility, but to transfer concentrated loads in the joints for elements with a large thickness. Such joints strength and endurance depend on the stress concentration level around the joints holes. In these joints, the stresses redistribution occurs, which leads to the structure strength change. The CM are reinforced with metal foil or high-strength films at the hole zones for the stress concentration reducing. The reinforcement using allows to reduce the structural weight without sacrificing performance. In turn, the weight reduction allows to increase the structural efficiency under dynamic loads.

The most important parameters that determine the type and structural features of mechanical joints are:

1) mechanical properties of the materials to be joined and additional reinforcement;

2) number and arrangement of fasteners;

3) holes dimensions and the distance from the holes to the joined parts edges;

4) joined layers' number in the structure.

Close-set dead joints are used to transfer a distributed load; dead combined joints are used to transfer a concentrated load with its redistribution. In this case, the most important parameters that determine the type and structural features of the joints are:

1) the ratio of the joined elements thickness;

2) the adhesive layer length and rigidity;

3) the adhesive layer physical and mechanical properties;

4) the shear moduli ratio of adhesive and mechanical joints.

The CM's ability to load redistribution during its operation should be taken into account when designing structures and joints. The following factors should also be considered:

1) the mechanical joints tightness during the part manufacture and operation;

2) threaded fasteners clamping forces;

3) dimensions accuracy, shape and location for fasteners holes;

4) CM properties changes during operation, that related with temperature and humidity changes, etc.

Taking into account the listed factors, it is necessary for correct designing process providing:

1) joint prototypes production and the full part for them testing purpose at the early possible designing stages;

2) part pilot series production, their checkout operation and making changes to the manufactural documentation.

5.2 Bonded joints

Close-set dead joints are obtained by using bonding adhesives that have an adhesion high level to the joined materials. Synthetic polymers are used as bonding adhesives. Structurally, the adhesive bead is made with an overlap with the possible to set additional reinforcement plates. The adhesive bead structure is shown in Figure 5.2.

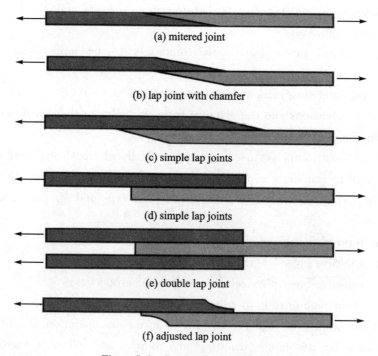

(a) mitered joint

(b) lap joint with chamfer

(c) simple lap joints

(d) simple lap joints

(e) double lap joint

(f) adjusted lap joint

Figure 5.2   Bonded joints structure

The manufactural bonding process includes the following steps:

1) surfaces preparation in order to ensure high glue adhesion (can have long duration and several stages);

2) applying a bonding adhesives layer to one or some glued surfaces;

3) bonded surfaces pressing to each other in ordered temperature and time conditions.

Advantages of bonded joints are:

1) high strength;

2) the ability to join parts that have a very small thickness;

3) the ability to join materials with different physical and mechanical properties;

4) the ability to join parts with different thicknesses;

5) no effect of the strength reducing of the joined parts;

6) the ability to ensure the leak-tightness and to exclude the corrosion;

7) good heat, sound, and electrical insulating properties;

8) the ability of complex shapes and parts creating.

The disadvantages of bonded joints are:

1) properties instability during operation (temperature, humidity, environment composition);

2) bonding adhesives toxicity;

3) fire hazard;

4) the manufactural equipment and the glued surfaces preparation process complexity;

5) insufficient joints efficiency for glued parts with high rigidity.

5.3   Molding joints

Molded joints are similar to bonded joints. The molded joints process creation is called molding-on or contact molding. These joints are widely used in the fiberglass (or other CM) large-sized parts manufacture. Molding-on is applying a filler layer (for example, fiberglass) impregnated with a binder at the parts butt. A strong joint is formed after the binder cures. Contact molding differs from molding-on in the operations sequence: first, a new filler layer is separately molded and impregnated with a binder, and after curing, a new layer is bonded to a part with the same matrix material.

5.4   Welded joints

Welded joints are close-set and are created by heating or by chemical interaction of the welded parts materials. In this case, there is a boundary disappearance between the parts materials in the joint zone due to their partial or complete mixing. In the case of polymer CM welding, the weld strength depends on the macromolecules size, shape and orientation. Parts from the same materials are most often welded. Welded parts material heating provides the material molecules interdiffusion in the welding parts zone. The same effect can be achieved by the solvent application to the welded surfaces. Diffusion welding is used for joining elastoplastic and thermoplastic parts. The main disadvantage of such joints is the material structure change in the weld zone.

Chemical welding differs from diffusion welding in that instead of a solvent, a chemical additive reagent is used, which starts a chemical reaction with the welded part material without the independent continuous phase formation. Chemical welding is used to join parts that are made of thermoplastics and thermosets with cross-links between molecules.

5.5　Threaded joints

CM part threaded joints are made by cutting threads on the joined parts mating faces, or by means of metal fasteners. The thread structure (its diameter and profile) for polymer CM mating parts differ from the standard thread types. As a rule, there are two ways:

1) a polymer CM part are joined with a metal part;

2) two polymer CM parts with a metal subpart with two threaded sections that placed between them.

Considering that the metal can withstand higher shear loads than polymer composites, the threads have an asymmetrical profile. Table 5.1 shows the threads profiles that are used for CM joints, their advantages and disadvantages.

Table 5.1　The threads profiles that are used for CM joints

Thread profile	Advantages	Disadvantages
V thread	Easy to manufacture and inspect, large shear area	There are radial stresses under load, stress concentration in sharp corners
Knuckle thread	Low stress concentration, large shear area	There are radial stresses under load, manufacturing and inspection complexity
Flat thread	No radial stress under load, easy to manufacture and inspect	Low shear area, low strength because flute is present, manufacturing complexity
Buttress thread	No radial stress under load, large shear area	There is stress concentration in sharp corners, manufacturing complexity

The thread joints strength is limited by the polymer binder low shear strength. There are two ways to increase the strength of such joints: the special structural solutions using and CM parts manufactural process adjusting.

Structural solutions are made by taking into account the loads types. So, for example, when a CM shell is loaded with pressure from the inside, the thread made external on it, and the metal mating part has an internal thread [Figure 5.3 (a)]. And, conversely, when loaded from the outside, a metal part is made with an internal thread [Figure 5.3 (b)]. It is possible to use structures in which a CM part has both internal and external threads with different length of thread engagement. In this case, the joint strength increases significantly, but the metal part manufacturing becomes more complex.

An adjusting in the CM part manufactural process is possible in the next directions:

1) increasing the binder elasticity;

2) special reinforcement lay-up in the time of a part manufacturing, which provides for the shear loading support (this way increases the stresses up to 8 times);

3) gluing the threaded joint (after which it becomes dead), increasing the strength up to 30%.

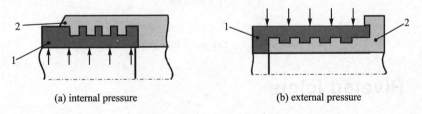

(a) internal pressure   (b) external pressure

1 —CM part (shell); 2—metallic part

Figure 5.3 Thread type for different pressure location

Metal fasteners (bolts and pins) are used to joint both thin-walled parts and flanges. In the case of thin-walled parts bolting (complete with nuts), their thickness and composition have the great importance. So, for parts with a thickness less than 1.5 mm, additional mending plates or ferrules are used. Relatively large sizes fasteners (diameter > 8 mm) are used to joint parts made of glass and organ plastics, and, on the contrary, small sizes are used for bioplastics and carbon plastics.

The number of fasteners and their location are very important factors. Since the holes, into which the fasteners are inserted, are stress concentrators, the distance between the holes and the part edge, as well as the distance between adjacent holes, are important. In this case, it is important to take into account: the load direction, its redistribution between the fasteners, the accuracy of the hole's shape and size, as well as their relative position accuracy.

The plug-bolt and plug-pin joints are also used for the parts butt-joint (Figure 5.4). The plugs, which are inserted into the hole's flange, operate as nuts. Flanges can have different structure, be made by various ways and strengthened by applying additional elements such as metal foil, boron films, etc. The plug holes can be located not in one, but in several rows in order to load distribution about the flange.

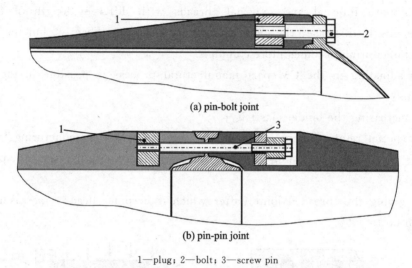

(a) pin-bolt joint

(b) pin-pin joint

1—plug; 2—bolt; 3—screw pin

Figure 5.4   The parts butt-joints

5.6 Riveted joints

Dead joints, obtained by rivets installation, are widely used for structural CM in different mechanic engineering areas. First of all, it is aircraft production. Riveted joints are made by the way of regularly located rivets rows (the so-called seams) that fixed thin-walled products. The most widely used rivets seam types are shown in Figure 5.5. The rivet is installed in pre-drilled holes in the joined parts, after that the closing rivet head is made. Before riveting, the joined parts are compressed.

Light alloy rivets are used to join CM parts. The riveting process is carried out without preheating. Rivets holes, as a rule, are drilled in the joined parts together, which makes it possible to prevent errors in their relative location. The most widely used items are rod rivets. The closing head is formed by a special impact tool (riveting hammers) or rolling tool (punches—rolling tools). If it is necessary to

(a) overlap joint

(b) butt-joint with one plate

(c) butt-joint with two plates

(d) stringers joints

Figure 5.5   Rivets seams types

ensure the riveted structure tightness, elastic gaskets made of rubber, metal, plastic are used. The best sealing is ensured by the special substance application— sealants, which after polymerization have high elastic properties and high adhesion to the part material and rivets.

In addition to rod rivets, different special rivets are used, examples of which are shown in Figure 5.6.

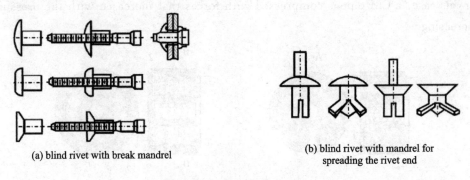

(a) blind rivet with break mandrel

(b) blind rivet with mandrel for
spreading the rivet end

Figure 5.6   Special rivets types

5.7   Self-locking joints

At the CM joints, the materials' strength is extinction due to additional stress concentrators near holes, grooves, threads. Above, in Figure 5.3, a design solution was shown that allows to increase the joint strength due to taking into account of the operational load direction. In the aircraft engineering, similar design solutions have found the application, that allows to take into account operational loads to increase the joint strength. These design solutions are based on the ensuring of compression stresses between the CM layers. The compression stresses should increase and, as a consequence, should increase joints strength and load-carrying capability with a load increasing on the joint.

The examples of self-locking joints are shown in Figure 5.7. As it can be seen in the figure, the jammed profiles are placed on conical mating surfaces. This solution allows evenly distributed load over the CM part thickness. The special profiling of the jammed surfaces allows evenly distributed load along the joint length.

Self-locking joints can be different with depending on the loading type and the supports structure. The detachable self-locking pipes joints that made of polymer CM, which are operated with internal pressure is shown in Figure 5. 7. Conical surfaces T-T have special jammed grooves, which can be spiral [Figure 5. 7 (a)] or annular [Figure 5. 7, (b)]. The inclination angle of the self-locking profile can be unchanged along the entire length, or changed with accordance to a required pass. The self-locking profiles can be located skew-symmetrically [Figure 5. 7 (a)] and symmetrically [Figure 5. 7 (b)]. The joint consists of two metal parts (2 and 3), which combined into single load-carrying structure by the single base. The load is evenly distributed between the metallic parts during operation under the pressure. In this case, a CM pipe is compressed with forces that increased with the pressure increasing.

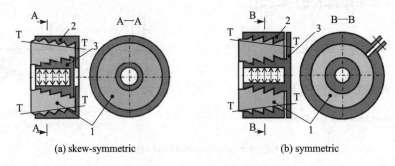

(a) skew-symmetric   (b) symmetric

1—CM part; 2, 3 —metallic parts

Figure 5. 7   Self-locking joints

5.8   Stitching and needle joints

The important factor is the uniform load distribution in the joint zone for thin-walled polymer CM parts. In cases when glued joints do not provide sufficient strength, but mechanical fastening with bolts or rivets significantly weakens the structure, stitching and needle joints can be used. The strength of such joints is close to the joined parts material strength, and their durability can exceed the bolted joints durability up to 18 times.

5.9   Combined joints

Combined ones include dead joints. They combine adhesive and mechanical joints (glue-riveted, glue-stitching, glue-needle, glue-bolt, glue-thread). Such joints can carry heavy loads, while simultaneously distributing them over the mating parts contacting area. They allow full using of the CM load-carrying capability, reducing the stress concentrators affect at the mechanical joints zones.

A combined joints special group is flanges molded into various shells and pipes, molded fittings and connecting ferrules. Such joints are designed to transfer high concentrated loads from metal parts to the CM parts.

Questions

(1) Briefly explain composite material parts of joint classification.

(2) What are the most important parameters that determine the type and structural features of mechanical joints?

(3) What are the factors that should also be considered when designing composite material structures and joints?

(4) Briefly describe bonded joints' features.

(5) Briefly describe molding joints' features.

(6) Briefly describe welded joints' features.

(7) Briefly describe threaded joints' features.

(8) Briefly describe riveted joints' features.

(9) Briefly describe self-locking joints' features.

(10) Briefly describe stitching and needle joints' features.

(11) Briefly describe combined joints' features.

Chapter 6 Polymeric Composite Materials Repair

Modern high-strength polymer CM, carbon-, glass- and organic- plastics, as well as honeycomb structures have great efficiency due to the decreasing in aircraft weight and the simplification of its production. Their application's increasing requires the development of new and improvement of existing repair technologies. In addition, honeycomb structures are more sensitive than other structural members to concentrated loads, and are often damaged by foreign objects. During the operation, hidden defects are determined that were not detected during their production.

So, here we will describe repairing approaches for CM parts for the aviation engineering. The main condition for aircraft repairs is the initial strength and aerodynamic performance restoration. This can be fully achieved by next manufacturing approach: repair part setting by prepress, film and pasty adhesives application with the required temperature and pressure for molding or gluing. The considered part repair approach makes it possible to do it from the same materials with the given reinforcement lay-up provision. In some cases, the described repair processes can be possible to use without removing parts from the aircraft, which will reduce costs and shorten the repair cycle.

6.1 Structural repair requirements

As noted above, during the aircraft operation, separate airframe parts damage is possible. During the repair process, primary structural members with huge damage should be removed. Damaged primary structural members replacing requires structural changes in the zone being repaired. At the same time, the primary structural member strength, stiffness and stability are changed, the structural mass are increased and there is a danger of the external surfaces outer shapes interruption. The listed parameters' significant deviations from their initial values can cause the repaired zones and the full aircraft failing. The repaired part stiffness changes are causes of the elastic deformation changing during flight and the outer geometric shape interruption, i. e. aircraft aerodynamic and flight performance decreasing. Both an increase and a decrease in stiffness cause the load redistribution, causing in the overloaded element durability decreasing.

An example of an incorrect repairing is the replacement of a carbon fiber skin part with a fiberglass one with non-observance of the skin reinforcement lay-up. The stiffer carbon-fiber skin takes as many times for the load as the elastic modulus of the carbon-fiber exceeds the elastic modulus of the fiberglass for the elastically tinseled panel. This load will be transferred at the skin joints border. The stresses in the joint between the skin and the repair plate (fiberglass) will increase by about the same number of times. Due to the significantly increased load repeating during the flight, the fiberglass repair panel can destroy at the joint with the rigid carbon fiber plate. From the given example it can be seen that the repaired element stiffness changes from the initial leads to the load redistribution. Overloaded elements can be destroyed during the aircraft operation. Therefore, it is necessary to ensure the materials' using (reinforcement and binder), similar to the skin being repaired, with the reinforcement lay-up obligatory keeping in the skin when repair plate installs.

Approaches and ways of ensuring the fulfillment of these repaired zones requirements on polymer CM units are reflected in the manufactural solutions. In addition, a very important point is to ensure the strength and rigidity equivalent of the repaired zone and the unit as a whole when performing repairs.

An airframe primary structural member's specific feature is that each of these elements supports a certain load at a given strain. The most common ones are tensile stresses. Based on Hooke's equation, it can be seen that the load supported by the primary structural element depends on the specific elongation ε and the product of the elasticity modulus E by the cross-sectional area of the element S. The loads equality (with the same elongation) taken by different primary structural elements will be only under the condition of equality of the $(E \times S)$ products, which is called the primary structural element stiffness. During the repair, the main condition for the repaired zone rigidity keeping is the observance of the equality $E \times S = E_P \times S_P$. Where E and S are respectively the elasticity modulus of the material and the damaged element area, and E_P and S_P are the same parameters for the repaired plate. Thus, it is possible to calculate the required cross-sectional area of the plate:

$$S_P = \frac{E}{E_P} S. \tag{6.1}$$

The element repaired zone can fail in one of five places when it is loaded with a tensile load:

1) on the skin weakness zone P_1;

2) by the compensate gasket glued into the repaired zone P_2;

3) at the compensate gasket with the skin boundary P_3;

4) along the repair plate with the skin overlap zone P_4;

5) on the repair plate P_5.

There is the simultaneous strength condition

$$P_1 = P_2 = P_3 = P_4 = P_5. \tag{6.2}$$

Compliance with this condition ensures sufficient repaired zone strength and a structural mass minimum increasing. However, it is very difficult to fulfill the exact observance of the simultaneous strength condition. Therefore, during repairing, the sufficient strength condition is expressed by inequalities:

$$P_2 \geqslant P_1, P_3 \geqslant P_1, P_4 \geqslant P_1, P_5 \geqslant P_1. \tag{6.3}$$

The case when P_2, P_3, P_4 and P_5 are much more than P_1 means excessive strength and mass increasing of the compensate gasket and repair plate, as well as the repair plate adhesion (overlap) zone increasing.

The breaking force P_1 is determined by the minimum cross-sectional area S_{weak} and the ultimate strength σ_b of the repaired skin material:

$$P_1 = S_{weak}\sigma_b = (S - \Delta S)\sigma_b, \tag{6.4}$$

where ΔS is the skin cross-sectional area reduction due to the defected section removal.

The breaking force P_2 is determined by

$$P_2 = S_{gask}\sigma_{bgask}, \tag{6.5}$$

where S_{gask} is the compensate gasket cross-sectional area that is glued in the repaired zone, σ_{bgask} is gasket ultimate strength.

From the condition (6.3) and equations (6.4), (6.5) it is possible to find the gasket ultimate strength that corresponds to the simultaneous strength condition:

$$\sigma_{bgask} = \frac{S - \Delta S}{S_{gask}}\sigma_b. \tag{6.6}$$

After that it is possible to select reinforcement lay-up.

P_3 is determined by:

$$P_3 = S_{gask-skin}\sigma_{bgask-skin}, \tag{6.7}$$

where $S_{gask-skin}$ is the compensate gasket and skin boundary cross-sectional area, $\sigma_{bgask-skin}$ is boundary material ultimate strength.

From the condition (6.3) and equations (6.4), (6.7) it is possible to find the boundary material ultimate strength that corresponds to the simultaneous strength condition:

$$\sigma_{bgask-skin} = \frac{S - \Delta S}{S_{gask-skin}}\sigma_b. \tag{6.8}$$

In practice, the material, used for joining (boundary between gasket and skin) is the prepreg binder from which the compensate gasket is made. The joint has

insufficiently shear strength when it is used as an adhesive when joining two cured
skins or when joining the cured skin and prepreg. In addition, this joint has
insufficient shear strength due to the commensurate joint length with the skin
thickness. To increase the shear strength, to ensure uniform load transfer, it is
preferable to obliquely joint between the repaired skin and a compensate plate. The
calculation of the oblique joint and its manufacturing specifics will be discussed
further.

P_4 is determined by:

$$P_4 = S_{over}\sigma'_{bplate\text{-}skin},\qquad(6.9)$$

where S_{over} is the overlap of repaired plate and skin cross-sectional area, $\sigma_{bplate\text{-}skin}$ is
overlap material ultimate strength.

From the condition (6.3) and equations (6.4), (6.9) it is possible to find the
overlap material ultimate strength that corresponds to the simultaneous strength
condition:

$$\sigma'_{bplate\text{-}skin}=\frac{S-\Delta S}{S_{over}}\sigma_b.\qquad(6.10)$$

To ensure a uniform load distribution on the adhesive, it is necessary to create
a serrated edge in the repair plate external layers, taking into account the
reinforcement lay-up.

P_5 is determined by:

$$P_5 = S_{plate}\sigma_{bplate},\qquad(6.11)$$

where S_{plate} is the repaired plate cross-sectional area, σ_{bplate} is plate material ultimate
strength.

From the condition (6.3) and equations (6.4), (6.11) it is possible to find the
repaired plate material ultimate strength that corresponds to the simultaneous
strength condition:

$$\sigma_{bplate}=\frac{S-\Delta S}{S_{plate}}\sigma_b.\qquad(6.12)$$

Having determined σ_{bplate} and knowing the material from which it will be made,
it is possible to determine the reinforcement lay-up. The existing reinforcement lay-up of the repaired skin with the possible application of additional layers in certain
directions is taken as a basis for choosing a reinforcement lay-up. The engineering
requirements for the airframe primary structural members repair are keeping at
least 90% of their initial strength. The skin and repaired plate oblique joint and the
plate external layers serrated edge application can make it possible to meet the
specified requirements when repairing units from the polymer CM.

6.2 Requirements for repaired structures

The first requirement for any structure repair is to restore the required strength and rigidity. It does not necessarily mean full structural strength restoration, as determined by the fibers break strength. Many aircraft units have limits for dynamic stiffness, durability, and fatigue, which are limited the design ultimate strength or ultimate strain to a value that significantly less than the material's break strength. The restored static strength should exceed the design ultimate strength, as close as possible to the structural initial material strength within the limits determined by other criteria. The repair process can be considered successful if 90% of the initial material strength is obtained. The restored stiffness should be close to the original material stiffness, because a significant change (decrease or increase) in it negatively affects the full structure.

The second requirement for structural repair (in the aviation engineering) is to ensure the aerodynamic surface quality. The aircraft structure has the aerodynamic smoothness criterion, which allows only little surface changes. For the most aircraft critical zones, surface irregularities of up to 0. 5 mm are allowed, although even more stringent limits can be for some specifics zones (such leading edges of the wing, air-intakes, etc.). Therefore, repair approaches must meet the specified requirements.

The third requirement is about structural weight. It is necessary to add that the weight increasing has been minimal after the repair process, especially this requirement is significant for control surfaces.

The next requirement is about environmental effects. The materials used for the polymer CM part repair, after their processing, must be compatible with environmental conditions. The polymer CM strength is mostly influenced by two environmental factors: temperature and moisture. The materials choice and manufactural processes are determined by the requirement to provide the required strength at temperature range $-60-+150$ ℃ in a humid atmosphere. In addition, it is necessary to take into account that real repairs will be carried out on materials that have been in the operation for some time and have some results of using (low strength, moisture absorption, etc.).

The materials and repair approaches must ensure the structure durability during the full aircraft cycling life.

In real life, it is needed to make a choice between damaged structure repairing at the given location of an aircraft (an airfield, an airport, etc.) or at an aircraft maintenance facility or at an aviation plant for carrying out repairs. This choice, i. e. the repair availability is determined by such factors as the ease of damaged unit removing, the aircraft forced downtime, the spare parts availability, the inspection after repair providing, the airworthiness requirements providing, etc.

6.3   Defects types

In the polymer CM parts manufacture, there can be a large number of factors that affect their performance. Many of these factors are interrelated. Failure, which determines the structural ultimate strength, is progressive and occurs due to material local defects and stress concentration. Partly this question was explained in the subsection 4. 4 of this book.

It is known that the polymer CM strength is affected by the voids presence or gas bubbles, the bond quality between the reinforcement and the polymer matrix, as well as surface defects that break the reinforcement integrity and continuity.

The void presence in polymer CM leads to the tensile and compression strength decreasing due to the structure cross-section area decreasing. The strength's dependence on the voids volume is calculated by the equation:

$$a = \sigma \left[1 - \left(\frac{V_v}{V} \right)^{2/3} \right], \qquad (6.13)$$

where a is the apparent material strength; σ is the theoretical material strength; V_v is the voids volume; V is the total material volume.

The total material volume is equal to the sum of the real solid substance volume and the voids volume. Experimental data shows that if the voids volume in fiberglass is 5%, the tensile strength decreases by 1. 15 times, and at case 10%, respectively, by 1. 25 times.

The bond strength between the fibers and the polymer matrix has great importance for the material mechanical properties.

It is necessary to take the chemical bonds between the polymer matrix and the fiber into account, as well as the fibers mechanical clamping, for the correct choosing of the polymer CM components and the manufactural process.

However, if the reinforcement specified moisture content is not provided during the prepregs production and their storage, then the CM strength can be significantly reduced and lead to failures in the operation, because the bond between fiber and matrix, skins and core is broken.

Surface defects can occur during the manufactural process, as well as during parts operation. This defects' type breaks the reinforcement integrity and continuity. Surface defects include the reinforcement stress wrinkle, the edges joint of one or more fibers layers with butt or overlap, macroscopic cracks, scratches, notches, etc.

The reinforcement stress wrinkles lead to the tensile and compression strength decreasing. The surface tears lead to the tensile strength decreasing, skin delamination, since exceeding the binder shear strength leads to cut fibers delamination, and subsequently to complete structure destruction.

It is very important to know in what conditions the materials will be operated. This is necessary to assess the effect of defects on the strength of the structure. It is possible that the same material, which has some defects, will operate satisfactorily under some conditions, and under other conditions, existing defects will be developed in it.

Cracks in the matrix are stress concentrators and affect the strength. They can be developed faster than delamination, and can have branching and lead to a sharp matrix destruction between the reinforcement layers.

Thus, the main destruction causes (full or partial) of polymer CM are:

1) spread of physical, mechanical and geometric parameters of the matrix and reinforcement;

2) insufficiently good adhesive and cohesive properties of the matrix and reinforcement, adhesive;

3) residual stresses in the matrix;

4) internal manufactural fiber microdefects, matrices at the "fiber-matrix" boundary (pores, cracks, delamination, cavities, stress wrinkles);

5) surface defects (scratches, stress wrinkles, cracks, prepreg overlaps, etc.).

Based on this, defects can be divided into two types:

1) defects that do not develop during polymer CM part operation;

2) defects that develop during operation and cause performance reducing, and sometimes lead to catastrophic events.

The most common defects in polymer CM structures are shown in Table 6.1.

Table 6. 1　Types of polymer CM structural defects and the reasons for their occurrence

Item number	Defect	Defect location	Defect specific	Defect sketch	Defect reason
1	Scratches:in matrix (binder) shallow scratch; in reinforcement (clothe, tape)	along full unit surface; along full unit surface	cracked part-through flaw; crack depth is smaller than 25% skin thickness		careless: transportation, storage, operation
2	Delamination	along full unit surface	the interlayer skin bond break		irregularities in the procedure (vacuum bag depressurization during the skin molding, poor-quality prepreg, film residuals, etc.)
3	Layer separation	skin —honeycomb core; skin — structural frame; structural frame— honeycomb core	the adhesive integrity break	delaminations	irregularities in the unit assembly (gluing) procedure, unacceptably high acoustic loads, moisture absorption
4	Bent	skin —honeycomb core	skin and honey-comb core strain		careless transport-ation, storage, me-chanical damage due to collision with foreign objects
5	Hole:one-side thru hole	along full unit surface; along full unit surface	local break of single skin and honeycomb core; local break of both skins and honey-comb core		careless transport-ation, storage, mecha-nical damage due to collision with foreign objects
6	Crack	along full unit surface	target-through skin integrity break		uneven stresses, unacceptably high loads

6.4　Repairing processes

Repairs at an aircraft maintenance facility or at an aviation plant are aimed at restoring the unit operational performance. To ensure a high-quality repair, it is necessary to carry out a set of preparatory processes to ensure the temperature and humidity conditions in the environment (temperature not lower than $+18$ ℃, relative humidity not higher than 75%).

Before repairing, it is necessary to:

1) determine the damage zone;

2) determine the damage boundary;

3) determine the skin thickness, its composition and the reinforcement type in the repair zone;

4) select appropriate repairing process, equipment, tools, materials, etc.

Before carrying out repairs, the repair zone must be cleaned of contamination. The repair plate installation on the zone being repaired can be performed in two ways:

1) gluing pre-made patches;

2) molding patches from prepreg in skin cut with partial replacement (if necessary) of the honeycomb core.

The second way is more preferable, because it allows to restore up to 90% of the initial strength.

All defects on repaired unit, as well as the repair operations, should be noted to the unit technical datasheet with an indication of the defect approximate shape, its type, dimensions and distance to the unit edges. The notes are saved during the full unit service life.

The repairing units' manufactural process begins with the marking defective zones. Then, the unit repaired zone is marked. Carrying out further repairing operations depends on the defect type.

6.4.1　Scratches repair

Shallow scratches repair (depth less than 25% of the skin thickness) is carried out using the following processes:

1) The paint coating is removed by sanding from the repair zone according to the marking.

2) The zone is processed with fine emery cloth for the scratch full length and half of its depth.

3) The scratch is sanded with emery cloth to the full depth. Then dust is removed from the defected zone with a clean dry brush.

4) 1—2 layers of prepreg are made and glued (molded) (it is possible to apply glass cloth impregnated with glue without filler), depending on the scratch depth.

The process of prepreg layers molding will be considered later when describing the process of repairing skin delamination from honeycomb core and holes in units with partial skin replacement.

Repairing of deep scratches is the same as repairing cracks.

6.4.2   Delamination repair

The repair process of skin delamination depends on its location. Delamination can be along the perimeter or at the inner zone skin.

For the delamination along the skin perimeter repair, it is needed that:

1) the defect is cleaned from the previous binder with emery cloth;

2) glue or binder is sprinkled with a pressure gun, and the repair zone is tightly compressed by hand;

3) excess binder or glue is removed with a cloth soaked in acetone;

4) the appropriate mode of the binder curing is carried out;

5) a package is collected from a release film (fluoroplastic, polypropylene), a heater, a thermocouple, a tsulaga, a heat insulator. On the opposite side (to the skin) porous rubber and a metal plate are on the top of it. Clamps with calibrated tightening are installed.

The inner zone skin delamination repair is carried out by the following ways:

1) Holes are drilled in the repair zone (Figure 6.1).

2) Rivet nuts are set in the holes, previously degreased in acetone.

3) The screw and rivet nut are coated with glue. The screw length should be less than the honeycomb height in the repair zone.

4) Glue is injected through the holes in the rivet nut and the screw is set.

5) Glue curing mode is carried out.

6.4.3   Layer separation repair

In polymer CM three-layer structures with honeycomb core, the following layer separations are possible (see Table 6.1):

1) the skin layer separation from the honeycomb core;

2) the honeycomb core layer separation from the structural frame;

3) the skin layer separation from the structural frame.

Manufactural approaches for these layer separations repair differ from each other. However, a common manufactural operation preceding any type of such repair is the moisture removal from honeycomb structures.

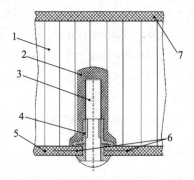

1—honeycomb core; 2—glue; 3—screw; 4—rivet nut;
5—delaminated skin; 6—delamination zone; 7—non-damaged skin

Figure 6. 1   Delamination repair with rivet nut

1. The moisture removal from honeycomb structures

Moisture accumulates in the zones where there are mechanical failures, as well as in the zones near the attachment fittings and the ribs with the spar joints. After unit moisture inspection, in the zones where it was detected, as well as in the zones of mechanical failure and layer separation, a moisture removal process should be done.

Approaches and equipment do not differ from each other regardless of the zone in which moisture is removed (skin — honeycomb core; structural frame — honeycomb core). When moisture is removed from the "structural frame — honeycomb" zone, access holes in the frame parts are used. When moisture is removed from the "skin—honeycomb core" zone, holes are drilled in the staggered order. Nipples are installed at the access holes (Figure 6. 2) for the vacuum system. The nipples are joined with sealant or rubber O-rings. The nipples can be made from transparent materials for visual inspection of the moisture removal process.

The repaired unit is placed in a clothes drainage layers and a vacuum bag is assembled. The assembled package is placed in a heat chamber or on a heating system that installed below. The vacuum about of 0. 03 MPa and temperature of (90 ± 5) ℃ can be provided into the bag. The repaired unit is maintained under vacuum and heating for about 6—8 hours. After that, the unit is cooled to 40 ℃. The unit is re-inspected for moisture. If moisture is detected again, the moisture evaporation is repeated. If it is absent, it is possible to start unit repairing.

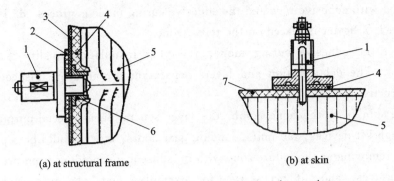

(a) at structural frame   (b) at skin

1—nipple; 2—rubber seal; 3—spar web; 4—sealant or glue;

5 —honeycomb core; 6—anchor nut; 7—skin

Figure 6. 2   Nipples arrangement for the moisture removal

2. Layer separation repair in "skin —honeycomb core" zone

Layer separation in "skin —honeycomb core" zone at units that do not have acoustic and vibration loads and do not have a special purpose in operation is repaired by sprinkling glue into the defect and installing plugs made of aluminum (if the fiberglass skin) or titanium alloys (Figure 6. 3). Holes for glue injection and installation of plugs are drilled in a staggered order. After holes drilling, dust and chips are removed from the processing zone with a vacuum cleaner.

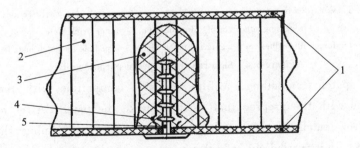

1—skins; 2—honeycomb core; 3—glue;

4—access holes in core webs for glue transferring; 5—screw plug

Figure 6. 3   Layer separation repair by screw plug

For repaired units that are at the heated engine gases zones, the adhesives with increased heat resistance are applied.

For repairs, paste-like adhesives are mainly used. They are highly viscous, and special syringes must be used to pump them into the repaired zone. The glue injection stops when there is a sharp increase in the injection pressure or when glue appears from access holes.

Immediately after glue applying, a plug is installed into the hole. The excess adhesive is removed with a cotton napkin dipped in acetone. The plugs' heads are supported with adhesive tape and the adhesive curing mode is processed. If heating is required, a heater is placed on the repair zone.

After the adhesive curing mode, the plug installation quality is visually inspected. The delamination and layers separation presence are inspected by a nondestructive tester.

The skin from the honeycomb core layer separation at special-purpose units (wing high-lift devices, tail units, landing gear doors, etc.) should be repaired by the skin removing at the failure zone with its subsequent molding from prepregs or film adhesives, and, if necessary (for example, metal honeycomb corrosion damage), replacing the honeycomb core section. The repairing process of such failure skin is rather complicated (Figure 6.4).

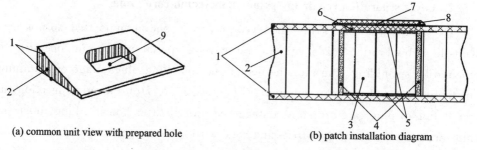

(a) common unit view with prepared hole   (b) patch installation diagram

1—skins; 2—non-damaged honeycomb core; 3—honeycomb core patch section;

4—foaming adhesive; 5—film adhesive; 6—inner patch plate;

7—reinforced film adhesive; 8—external patch plate; 9—cut hole for patch installation

Figure 6.4   Skin repair for special-purpose units

Before patch's installing, a film adhesive compatible with the binder is impregnated with the prepreg that used to make the patch. After molding and installing the patches, the patches curing process and gluing them to the honeycomb core is carried out at high temperature and overpressure.

3. Layer separation repair in "structural frame —honeycomb core" zone

For such a failure repair, two holes are drilled in the repaired zone opposite the ends (Figure 6.5). The pressure on the drill (feed) should be minimal when the holes are drilling to avoid breaking the part inner layers.

The chip is removed from the repaired unit by a vacuum cleaner and the holes are closed with adhesive tape. It is the picked hole in one adhesive tape with a needle to flow air when the adhesive injects. The repaired unit should set so that the repaired part will be at upper horizontal position to prevent adhesive spontaneous leakage. The second adhesive tape is also the picked hole and adhesive is injected in it. The adhesive is sprinkled until the adhesive starts to spontaneously flow out of the opposite hole. The adhesive tape is removed from the hole and a titanium or steel screw is screwed into one hole. The repaired zone is covered with 1—2 fiberglass layers with cold curing adhesive to ensure the adhesive joint tightness. After completing the specified work, the adhesive curing mode is performed.

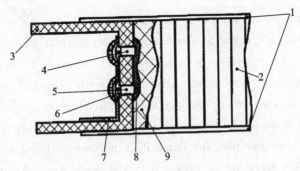

1—skins; 2—honeycomb core; 3—rear spar; 4—hole for the excess adhesive; 5—injection adhesive hole;
6—screw plug; 7—impregnated adhesive cloth; 8—"new" injected adhesive; 9—"past" adhesive

Figure 6.5   Layer separation repair for "structural frame—honeycomb core" zone

4. Layer separation repair in "skin—structural frame" zone

This type of failure is of two types:

1) the separation with access from the outer side and the possibility adhesive injection into the clearance space between the skin and the structural frame;

2) without external access to the skin and the frame separation.

The repairing of skin separation from the structural frame with access to the outer side is performed as follows. Initially, it is necessary to make sure that the separation has not reached the honeycomb filler, otherwise, before carrying out the process, the moisture absence in the repaired unit is inspected. The repairing is carried out by injecting adhesive into the clearance space by using a thin plate or by sprinkling adhesive in that case.

The second failure type is repaired by removing a skin part and restoring it by simultaneous molding from prepreg and gluing to the frame, as it is described above.

6.4.4   Cracks repair

The cracks repair way depends on the repaired unit requirements. If the requirements are rigid, then it is necessary to remove a skin part and restore it by simultaneous molding from prepreg and gluing to the honeycomb core and the structural frame.

1. Preparation process

If it is possible to install a patch, the repairing is carried out as follows. Initially, the crack location is determined. There are two variants:

1) the crack is at the skin edge, but does not presents or presents on the honeycomb core;

2) the crack does not present at the skin edge.

In the case that a crack is at the honeycomb core, it is possible that moisture has got into it. The moisture is inspected. If it is present, in this case, skin delamination from the honeycomb core additionally checked due to moisture freezing. For such case, before crack repairing, moisture is removed from the unit and delamination is repaired. The crack end is drilled to the full skin thickness till the adhesive. In the case that the crack does not reach the skin edge, a hole is drilled and then the moisture is inspected in the crack zone. After drilling holes, moisture inspection and its removing, the patch is glued.

2. Patch installation process

The most preferred patch shape is circle or ellipse. The patch overlap is determined depending on the skin thickness. The size ratio of the failure zone for the patch should be not more than 1:2.

The finished patch is cut from a sheet of the same type of material as the skin, and of equal thickness. The material from which the patch is cut must have a "sacrificial layer" on the surface to be glued to exclude the surface contamination possibility and the need for sanding before gluing. A patch with a thickness of more than 0.5 mm should have chamfer at an angle of 3° to 10°. The patch cut and chamfer line are preliminarily specially processed with an appropriate binder or adhesive and heat-treated for the moisture penetration excluding into the patch through the cut fibers ends, which were formed during machining. A phenol rubber adhesive layer is applied to the patch glued surface (the "sacrificial layer" is removed beforehand) when for patch gluing pasty epoxy adhesives are used. The adhesive is applied to the patch in three layers with an open time:

1) after the 1st layer: 30—60 minutes at 18—30 ℃;

2) after the 2nd layer: 30—60 minutes at 18—30 ℃;

3) after the 3rd layer: 3 hours at 18—30 ℃.

The adhesive layer is applied evenly, without bubbles.

After the total adhesive open time, it is cured at the temperature of 125 ± 5 ℃ for 4 hours. The temperature rise rate is no more than 2 ℃ per minute. Before patch gluing, the glue sublayer must be sanded without degreasing, sanding products are removed. Patches covered with a one-coat adhesive can be stored for 1 year at regular temperature when packed in film. One-coat adhesive contamination is not allowed.

Before the patch gluing, the repaired unit surface is sanded to a uniform roughness. An adhesive tape is applied along the failure zone perimeter to protect it from adhesive rundown. Paste adhesive is prepared or blanks are cut out of the adhesive film. Paste adhesive or an adhesive film must be applied, and the patch is installed in accordance with the marks.

Repair package, heater, heat insulator are set on the patch according to Figure 6. 6.

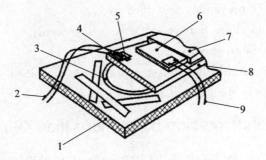

1—repaired skin; 2—thermopile; 3—patch boundary; 4—release film (fluoroplastic or polypropylene);
5—adhesive tape; 6—heater; 7—heat insulator; 8—aluminum alloy draw sheet; 9—heater wires

Figure 6. 6   Repair package scheme for the patch installation

In the repair package, various layers are used with the following functional purposes:

Release film is a separating layer between the prepreg and the draw sheet, ensures the quality of the patch external surface. The release film heat resistance is not less than 20 ℃ higher than the highest curing temperature.

Aluminum alloy draw sheet is prevented with a sharp temperature difference in the polymer CM skin at the heater edge, i. e. prevents the skin and patch warping, provides the patch surface high aerodynamic quality.

Heat insulator is limited by the heat transfer (i. e. heat loss) from the repair zone, protects the vacuum bag from thermal damage.

Thermocouples are placed on the patch edges and two film layers are placed over it, and then applied with adhesive tape.

After repair package installation, the adhesive is cured according to the ordered modes. After curing, the patch quality is inspected.

6.4.5 Dents, break-through and one-side holes smaller than 40 mm repair

These failures are repaired by adhesive applying at the failure zone and patch installation. Foaming, epoxy adhesives can be used, which contain various binders (microballoons, micro-adamantine spars, chopped fiberglass). The repair process is carried out as follows:

1) initially, the skin is machined in the failure zone: creases, bends, cracks are cut and sharp edges in the corners are chamfered;

2) the moisture inspected in the failure zone, and if moisture is present, it is removed by using the previously described ways;

3) the aluminum alloys honeycomb core is degreased;

4) the adhesive is prepared;

5) the patch is installed by using the previously described ways;

6) the final step is the patch quality inspection.

6.4.6 Dents and one-side holes more than 40 mm repair

This failure type is repaired by removing the skin and honeycomb core sections, gluing a honeycomb core section and gluing or molding a skin patch.

As it is noted earlier, honeycomb core made from aluminum alloys, as well as polymers and fiberglass, are used in the units' structure. The general condition for the honeycomb core gluing is providing heating of the repaired zone. For aluminum honeycomb core repairing, heating can be performed from the side opposite to the failure. But, the skin should not overheated under the heater. For polymers and fiberglass honeycomb cores, heating is carried out on both sides of the repaired units.

The cold-curing adhesive is applied with heating up to 70 ℃ along 1 hour for units repair that operate at temperatures up to 80 ℃. The foaming adhesive is applied with heating that not lower than 120 ℃ for units that operate at temperatures up to 150 ℃.

The clothes are used when pasty adhesives are used. The fiberglass clothes are used when film adhesive is used. That is, in both cases, it is needed preliminarily to make an inlay, with the help of which the honeycomb core and the insert will be joined at the failure zone.

Next way for such repair (see Figure 6.7):

1) the adhesive is prepared;

2) the adhesive is applied to the insert and honeycomb core;

3) the insert is installed to the prepared hole;

4) the insert is fixed by adhesive tape from free sides;

5) the adhesive curing process.

Thus, after the gluing operation for the honeycomb core, it is possible to perform the patch gluing or molding operation. The way of creation and gluing (molding) patches was considered earlier.

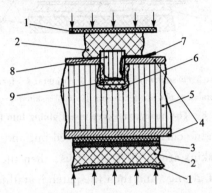

1—vacuum (air) bag; 2—heat insulator; 3—heater; 4—skins; 5—honeycomb core;

6—foaming adhesive; 7—thermopile; 8—honeycomb core insert; 9—inlay

Figure 6. 7   The insert installation and the patch applying

6.4.7   Thru hole repair

The greatest interest from the point of view for original engineering solutions is the repair operation of the honeycomb insert gluing, as holes in the skins (after the failure zone machining) can be of the same or different sizes.

The insert structural height is equal to the removed honeycomb core height when the inner patch is formed. Then, the honeycomb insert degreasing operation is performed (in the case of an aluminum honeycomb core using) and a foaming paste or foaming film adhesive is applied onto the insert side surface. The insert is installed into the hole and the repair package is created (see Figure 6. 8). Thermopiles should be installed one at each insert end and one directly into the adhesive. The fluorocarbon film is arranged on the repaired zone bottom surface, above the film there is an aluminum alloy plate (the plate and the film are fixed with adhesive tape). The plate and film are larger than the honeycomb insert. The fluorocarbon film is arranged on the failure zone upper surface, above the film there is aluminum foil or an aluminum alloy sheet. Heaters and repair packages are installed on both sides and a pressure of about 0. 01—0. 015 MPa is created. If the aluminum alloy core height in the repair zone is less than 100 mm (for the CM core is less than 20 mm), one heater is used, arranged from below. The pressure is created by using vacuum or air bags.

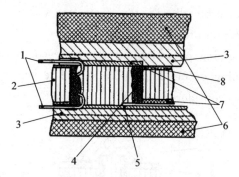

1—thermopiles; 2—base honeycomb core; 3—heaters; 4—foaming adhesive;
5—aluminum alloy plate; 6—heat insulator; 7—skins; 8—cover plate

Figure 6.8   The honeycomb core insert gluing into thru hole

Subsequently, the glue curing mode is carried out and the skin is repaired.

If the holes in the skins are of different sizes, then the repairing is carried out by the honeycomb insert gluing, and then the patch installing from the smaller hole side and the patch gluing (molding) from the larger hole side (see Figure 6.9).

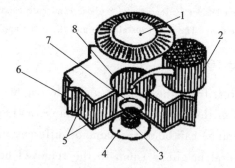

1—upper skin patch; 2—honeycomb core insert; 3—clothes inlay (its thickness is equal to the skin thickness);
4—bottom skin patch; 5—skins; 6—honeycomb core; 7—smaller skin hole; 8—larger skin hole

Figure 6.9   Thru hole repair

6.4.8   Units tips repair

Units tips are: winglet, fin tip, blade tip, stabilizer tip, etc. They can be made in various ways: with a tip stringer, without a stringer, with single-piece skin. These units' repair is very difficult due to the need of the honeycomb insert machining. Preferably, the honeycomb insert is made without allowances and then glued at the repair zone. The stringer repair and the insert gluing are done with cold curing adhesives using. In this case, the repair operations sequence will be shown in Figure 6.10.

The unit failure section is initially removed. The cut is made as simple as

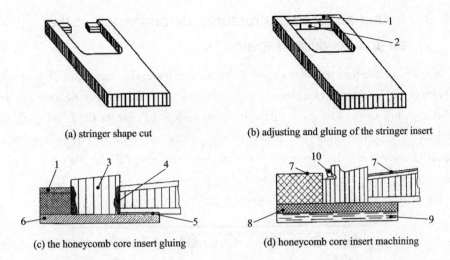

(a) stringer shape cut   (b) adjusting and gluing of the stringer insert

(c) the honeycomb core insert gluing   (d) honeycomb core insert machining

1 —stringer splice plate; 2—stringer insert; 3—honeycomb core insert; 4—adhesive; 5—release film;
6—temporary plate; 7—adhesive tape; 8—rubber plate; 9—plywood plate; 10—blade cutter

Figure 6.10   Units tips with a stringer repair

possible to the honeycomb insert and patches break-in procedure. The stringer undamaged ends are clean out on both sides and cut them. The stringer missing part is made from the same structural material. The stringer glued surfaces are covered with adhesive, followed by heat treatment at 125 ℃ to ensure the adhesive bond strength.

Then proceed to the honeycomb can insert creation. A temporary plate with a size larger than the patch gluing zone is used to ensure the honeycomb insert gluing. If there are irregularities on the skin external surface, the inlay should compensate them (it consists of several layers). The pre-prepared pasty adhesive is applied to the honeycomb core glued surfaces in the unit and to the stringer, as well as to the stringer insert and splice plate. The honeycomb insert is installed into the hole and then the stringer insert and the splice plate are installed. On the base side, a release film is fixed with adhesive tape, then a temporary plate is installed. The glue curing process is carried out. The honeycomb outsize is then removed. A rubber sheet together with plywood plate are arranged on its surface from the base side to ensure processed core fixation. The plywood plate is pressed to the unit and fixed with adhesive tape so that it prevents the core deformation under the cutting forces. The honeycomb outsize is removed. After honeycomb core adjusting, the repair package is disassembled and the skin restore operation is performed.

6.4.9　Sound-absorbing structures delamination in the air-intake ducts repair

Sound-absorbing structures are CM honeycomb panels, the inner skin of which is perforated. During these structure operations, the skins may delaminate from the honeycomb core. The most difficult repair task is to restore the inner perforated skin from the honeycomb core delamination, since some of the holes are covered with adhesive during the structure manufacture, and also during the operation, the sound-absorbing panel becomes dirty through the holes.

Repairing is carried out by adhesive injection according to Figure 6.11.

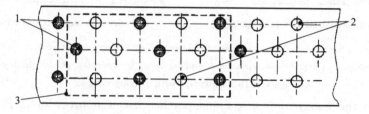

1—holes without adhesive; 2—holes for the adhesive injection; 3—repair zone outline

Figure 6.11　Adhesive injection diagram for the sound-absorbing structures

The repairing zone is cleaned with a high vacuum cleaner or a compressed air jet. Then adhesive is injected into the failure zone holes. The injection is done before the pressure in the syringe rises. After that, it is necessary to heat the repairing zone to a temperature of about 60—70 ℃ until the adhesive is completely cured. The holes that were injected with adhesive during repairing should be cleaned by an awl or drill using.

6.4.10　Manufactural solutions for the repair quality increasing

1. Machining process

The adhesive bond creation during the skin delamination repair, the patch installation, etc. require mandatory failure zone machining procedures, both to remove micro- and macro-stress concentrators, and to prepare the surface for bonding. The analysis and the researches showed that the best results were obtained when machining with a countersink drill. The surface roughness of the polymer CM depends on the feed and the cutting speed. The cutting speed has less effect than the feed and an abrasive grit. The microroughness decreasing at cutting speeds of 15—18 m · s^{-1} and feeds of more than 300—400 mm · min^{-1} can be dependent on the binder thermomechanical destruction. A similar event was found at low cutting speeds up to 3 m · s^{-1}, which occurs due to the fiber elastic behavior at such speeds, in opposite to brittle break at high speeds. Figure 6.12 shows the feed and speed effect on the surface roughness of the carbon fiber reinforced polymer KMU-4E (Russia).

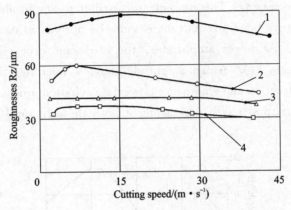

1—abrasive grit si800/630 and feed is 390 mm · min^{-1};2—abrasive grit is 800/630 and feed is 30 mm · min^{-1};

3—abrasive grit is 500/400 and feed is 390 mm · min^{-1};4—abrasive grit is 500/400 and feed is 30 mm · min^{-1}

Figure 6.12   The feed and speed cutting effect on the surface roughness of the carbon fiber reinforced polymer KMU-4E (Russia)

The adhesive bond strength depends not only on the surface roughness, but also on the capillary-active groups' number on it, which are formed due to the polymer mechanochemical destruction. The active groups' number depends on the cutting speed. The active groups concentration increasing at high and low cutting speeds are confirmed by the polymer binder destruction degree increasing. The acetone using for the degreasing leads to the active groups number decreasing on the surface by 20%—30%. The cutting speed and surface preparation way effect on the active groups number is shown in Figure 6.13.

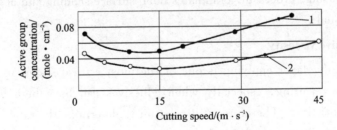

1—without surface preparation；2—acetone degreasing surface preparation

Figure 6.13   The speed cutting and surface preparation way effect on the active group concentration

2. Ultrasonic process

Improving the adhesive bond quality during repairing can be achieved by using ultrasonic processing. It was found that, as a result of the high-frequency mechanical vibrations application to the adhesive bond, the adhesion area is increased. It is possible due to the acoustic capillary effect, which is the phenomenon of the liquid penetration depth and rate increasing under the ultrasonic action. The ultrasonic pressure increases with a surface tension increasing and a liquid viscosity decreasing. The concentrator oscillation amplitude has the greatest influence on adhesive bond strength increasing. Its optimal values are in the range of 4—7 microns. At lower amplitudes, the ultrasonic processing time sharply increases (Figure 6. 14). At an amplitude above 7 microns, the reinforcement fibers partial destruction and emergence powerful cavitation process occur, which in total leads to a sharp strength decreasing of both the structural material and the adhesive bond.

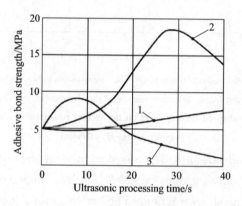

1—amplitude 3 μm, 2—amplitude 5 μm; 3—amplitude 15 μm

Figure 6. 14   Adhesive bond strength dependence on the ultrasonic processing time

After high-frequency vibrations' processing, it is possible to reduce the pressure during the adhesive bond polymerization, because it is necessary only in the case of the bonded surfaces with some non-flatness relative to each other. In addition, ultrasonic processing eliminates bond surface sanding and degreasing with flammable liquids.

3. Adhesive viscosity

The adhesive layer applying is one of the most critical stages in the polymer CM structure repairing, especially when adhesives can have high viscosity that grows in a short time. Therefore, it is relevant to determine the rational processing parameters for adhesive-mechanical joints.

The adhesive is a visco-plastic liquid. The adhesive VK-27 (Russia) plastic dynamic viscosity change (Figure 6. 15) is described by the equation

$$\eta = \eta_0 e^{kt} - 490e^{1.94-10^{-4}t} \tag{6.14}$$

where η_0 is adhesive viscosity at the final of its preparation; k is tangent of line to the horizontal axis; t is pressure duration.

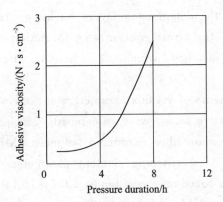

Figure 6.15　Adhesive VK-27 (Russia) dynamic viscosity change during its polymerization

Thus, it is possible to determine the rational polymer CM repairing process parameters of adhesive bond with optimal thickness and with given structural specifics. It is especially important when repairing is carrying out directly on an aircraft, when the high pressure creation is difficult due to the units' structural features and the repair conditions.

4. Stitching process

Failures such as delamination are most common in layered polymer CM. These reasons include non-adhesives, air macro-inclusions, release film parts, etc. If compression, shear, or bending loads occur in the failure zone, then the delamination can cause a local or general structure stiffness decreasing, a change in the external structure shapes during delaminated layers buckling, or the delamination zone increasing and total structure destruction.

One of the promising ways to structure repair in such case is the failure zone stitching process with threads impregnated with a binder or polymer CM pins (fully cured or in the prepreg state), which is created by pultrusion from impregnated unidirectional threads (Figure 6.16).

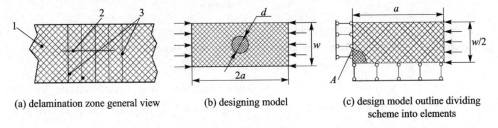

(a) delamination zone general view　　(b) designing model　　(c) design model outline dividing scheme into elements

1—polymer CM plate; 2—delamination zone; 3—stitching elements

Figure 6.16　Stitching elements installation diagram

The stitching process using provides to reduce the stress intensity factors at the delamination top, and also to significantly increase the critical compression load that corresponding to the delaminated layers buckling. However, the stitching process using leads to the stress concentrates formation in the zone where the stitching elements are installed, which, as is known, affects the structure load-carrying capacity.

The stitching elements of various parameters and their arrangement influence on the stress-concentrative factor were researched. The plate elastic parameters were taken from the carbon fiber reinforced polymer KMU – 4L (Russia) with layers ($0/\pm 45°/90°$). The stitching element parameters were taken from the unidirectional fiberglass based on epoxy binder EDT – 10 (Russia) and glass fibers from high-modulus glass VM-1 (Russia). The stress state was evaluated in the point A (Figure 6.16) by the stress-concentrative factor change calculation:

$$\frac{k_\sigma^e - k_\sigma^{e-1}}{k_\sigma^e} \times 100\%, \qquad (6.15)$$

where k_σ^e is stress-concentrative factor for $\eta = w/d$; k_σ^{e-1} is stress-concentrative factor for $\eta = \dfrac{w-d}{d}$.

The stress-concentrative factor change assessment was carried out for three stitching element diameters (1, 3 and 5 mm). The calculation results analysis, presented in Figure 6.17 as graphs, show that with the relative size of the $\eta > 6$, the change in stress-concentrative factor for all diameters does not exceed 2%.

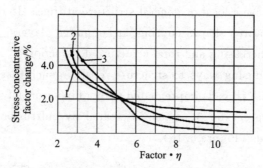

1—diameter 1 mm; 2—diameter 3 mm; 3—diameter 5 mm

Figure 6.17　The plate relative dimensions and the stitching element diameter influence on the stress-concentrative factor change

Thus, for polymer CM structure repairing that has delamination by this method, the stitching element hole can be required with the smallest possible size, taking into account the manufactural features of stitching. An anisotropy increasing in the factors between materials leads to the stress-concentration of factor. Therefore, for stitching elements, it is necessary to select materials whose elastic features only slightly differ from the elastic features of the structural material.

Questions

(1) What are the structural repairing requirements?

(2) What are the requirements for repaired structures?

(3) What are the main destructive causes (full or partial) of polymer composite materials?

(4) List the main types of the polymer composite materials' defects.

(5) Describe the process before repairing.

(6) Briefly describe scratches repairing process.

(7) Briefly describe delamination repairing process.

(8) Briefly describe layer separation repairing process.

(9) Briefly describe cracks repairing process.

(10) Briefly describe the repairing process of dents, break-through and one-side holes smaller than 40 mm.

(11) Briefly describe the repairing process of dents, break-through and one-side holes more than 40 mm.

(12) Briefly describe thru hole repairing process.

(13) Briefly describe units tips repairing process.

(14) Briefly describe sound-absorbing structures delamination in the air-intake ducts repairing process.

(15) What are the manufactural solutions for the increasing of repairing quality?

Chapter 7   Composite Materials Structures Safety

The maximum weight and economic efficiency of the polymer CM structures with ensuring their safety and reliability required level can be achieved with considering the following problems:

1) the rational combination choosing of the base materials various types and manufacturing processes with a polymer CM unit carrying-load structure;

2) the design and technology support for the polymer CM structure operational survivability.

7.1 General safety issues for composite materials structures

The materials selection criteria, design and technological solutions should be based on a closed-loop multidisciplinary approach to the polymer CM structures design to ensure their weight and economic efficiency. The approach should be based on an analytical way that simulates the multipara metric process of a new product creating that based on the key factors analysis.

Material selection criteria:

1) the level of failure rate and repair approaches in operation;

2) the manufactural processes (with taking into account specific defects) and quality inspection ways;

3) the part overall dimensions' restrictions;

4) the batch sizes and performance;

5) the regulatory and engineering support.

Also, there are many ways for the design solutions choosing that directly related to the design and manufactural process for polymer CM structure:

1) manufactural features;

2) metal-composite materials specifics;

3) holes and cutouts in a structure;

4) issues of operational damageability and survivability;

5) inspection and maintainability.

There is a scheme for this methods implementation. It is "Building Block" concept that is developed for the modern engineering for the certification basis creation.

7.1.1   Polymer composite material structure creation on the "Building Block" base

Weight and economic efficiency ensuring methods, which were developed over the past 100 years for aircraft metal parts, can be based on the deterministic properties of the structural materials, semi-finished products and technologies. For this approach, the carrying-load structure is developed independently of these factors. But, the material supplier is responsible for ensuring the materials properties that was determined during the designing process.

This approach is inapplicable to polymer CM structures. It does not allow to use total CM advantages. Especially it relates with the weight and economic efficiency. Engineering process in this case cannot be based on the fixed sequence such as " structure-material-technology ". All these three components are interdependent and are developed simultaneously. For the CM with given technical requirements set there are "material-technology-structure" variety combinations. Under these conditions, the "Building Block" concept can provide the safety and weight efficiency (Figure 7.1).

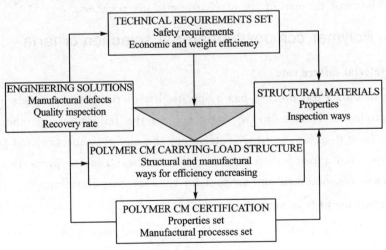

Figure 7.1   "Building Block" concept diagram

This approach takes into account the mutual properties influence of the structural materials, manufactural processes and carrying-load structure. It provides for the developer with significant flexibility in achieving target craft performance and, also, imposes the obligations and the responsibility on him about materials selection, its certification, manufactural processes selection and CM parts quality inspection in production.

This approach should be based on an analytical way that makes it possible to simulate a multi-parameter process of new part creation based on total key factors analysis. It should provide an assessment of the various carrying-load structure effectiveness at the earliest design stage (Figure 7. 2).

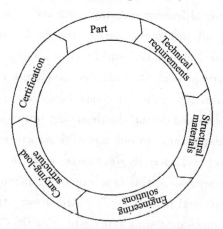

Figure 7. 2　Closed-loop feature of the "Building Block" concept

This analytical way should contain the corresponding databases and expert systems, which are used for structural variants simulation and analysis with the choice of the most rational of them according to given criteria.

7.1.2　Polymer composite material selection criteria

1. Material failure rate

Each manufactural process has a specific list of the base materials and semi-finished products. It has directly influence on the failure rate of the CM and structures from it. This, in turn, determines the quality inspection and testability requirements and repair efficiency, as well as the structure properties spread, which should also be taken into account for the materials choosing.

2. Repair methods in operation

At the base materials' stage selection for polymer CM structures, it is necessary to take into account the available repair processes. It is needed to provide parts and semi-finished products list that should be ordered for repair using. This list should take into account availability, storage period (which is especially important for prepregs) and recycling technology of these CM parts.

7.1.3   Criteria for selecting the manufactural process

1. Manufactural defects and quality inspection

The most important factor of determining the success in the development of new polymer CM structures is the developed theoretical manufactural process of accuracy's application in a real productive process. Different manufactural approaches have different sensitivity points that affect manufactural mistake: the human factor, an excessive number of manufactural conversions, the industrial equipment geometric instability in the heating process, the manufactural package instability in the pressure process, the binder shrinkage effect, etc.

This means that when choosing manufactural approaches, it is necessary to take into account the applied common quality inspection and the inspection of the most important CM parts zones (taking into account the tolerances inherent in the carrying-load structure). Otherwise, all the advantages of an optimized design process can be negated by errors in serial production.

2. Overall dimensions' limitations

Dimensional limitations are formed depending on the type of equipment and production accessories used. For vacuum molding and vacuum infusion, this is the size of the furnace inner chamber. For the resin transfer molding (RTM) technology, the limitations are determined by the capabilities of the punch lifting and moving within the production area, as well as the cost of the "die + punch" set. For the pultrusion, the limit is the extruded profile transverse dimensions, depending on the type of equipment.

3. Typical batch size and performance

It is necessary to have a clear meaning of the manufacturing cycle duration, taking into account more productive and less productive processes. For example, RTM technology, which is based on the same principle as vacuum infusion, provides higher productivity than the latter, but it is rational for relatively large batches, since the RTM equipment will be cost-effective much longer. Therefore, vacuum infusion can be effectively used for the manufacture of large-sized parts for relatively small batches, for example, main wing panels. But, high-performance approaches such as pultrusion and RTM for the parts manufacture have a smaller size, but are produced in a larger batch, for example, stringers, ribs, window frames.

4. Requirements for regulatory and engineering support for certification

This criterion influence on the manufactural approaches choice is mainly determined by the certification requirements for the created structural elements. The higher the critical component level is, the more detail and depth inspection of the materials and semi-finished products properties and the manufactural process stability may be required by the certification agency. Therefore, at the selection stage, it is necessary to ensure the regulatory and a certification base availability, which will be used to ensure aircraft type certification, as well as certification of serial production.

　　Considering that in modern time the computational approaches created for the polymer CM are not as reliable as for metals, mainly due to the destruction modes variety for possible damages, the "Building Block" approach has become traditional in the USA and Europe for the certification base development (Figure 7.3).

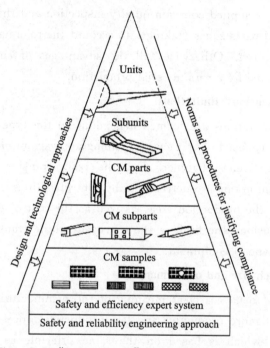

Figure 7.3 "Building Block" approach principle scheme

　　On its basis, the engineering approaches and their further applicational possibility are checked to substantiate certification requirements, which in current practice leads to an ever-increasing tests volume and the certification cost and timing increasing (such as the Airbus 380 and Boeing 787 programs). The important condition for overcoming these problems and ensuring the "Building Block" approach effectiveness is the methodology and the expert system in its basis. The methodology determines and justifies ways to achieve certification targets (primarily safety and reliability). The expert system ensures this methodology practical implementation in order to achieving the polymer CM maximum efficiency in aircraft structures. The computational and experimental researches development and implementation of the polymer CM structures safety and efficiency at each "Building Block" level is ensured by the rational application of design and technological approaches to increase weight efficiency and certification standards and procedures for justifying compliance.

7.1.4   Criteria for selecting the design solutions

There are traditional requirements for static and fatigue strength, as well as for aero elasticity. But there are additional criteria that can be directly related to the issues of materials science and manufacturing approaches of polymer CM structures.

1. Manufactural requirements

Aircraft structural elements differ significantly in their rational production approaches. Among the structural members, for example, it is possible to distinguish: thick-walled large-sized (spars and frames), thick-walled small-sized (rods and levers), integral (reinforced wing and fuselage skin panels), closed-loop (fin and stabilizer torsion box), envelops (skins), profile (floor beams), etc. For each of the listed structural member's types, the rational combination of basic materials, semi-finished products and manufactural approaches for their processing can be selected. For example: for large-sized, it is vacuum infusion; for thick-walled and small-sized, it is RTM; for profile, it is pultrusion, etc.

2. Metal-composite joints

It should be taken into account that the factors determine the features and limitations of collaboration of metals and CM when using combined metal-composite elements in a carrying-load structure. For example, it is recommended to use titanium instead of aluminum together with the CM according to their thermal-expansion coefficient. For a lightning protection system designing, it is necessary to take into account that a lightning strike in the area where metal parts are under the CM will lead to a polymer CM breakdown and the shell hole creation.

3. Holes and specific structural cut-outs

It is necessary to provide for a special reinforcement scheme with a special fiber lay-up angle for designing structural cut-outs in the polymer CM parts. Such scheme ensures a high-quality implementation of cutting or drilling processes. These processes are a complex procedure based on sequential deepening of the hole with the inner surface and hole edges state constant monitoring.

4. Damage tolerance

The design solutions applied for the aircraft elements should ensure that the damage per hour rate is low enough for the aircraft safe operation between routine inspections, or provide for reserve parts that stop the damage development due to the loads redistribution.

5. Operational damage rate

Operational damage of different types is for different airframe parts. For example, according to statistics, a lightning strike occurs mainly at the fuselage nose, wingtips and tail, but almost never occurs in the fuselage and wing central sections. The hail impact, as well as the tool fall during maintenance, is impossible on the fuselage and wing lower sections panels; the bird strike falls mainly on the wing leading edge. Thus, the dynamic effect of types from low-speed and high-speed impacts can be differentiated by structural parts and taken into account for the reinforcement schemes choosing, their thicknesses and structure.

6. Inspectability

Structural parts and zones most susceptible to the operational damage should be accessible for their condition inspection and monitoring by the maintenance staff, with taking into account their qualifications and inspection ways. These methods include: visual inspection, rapping (which is effective only in conditions of sufficient silence and qualified staff inspection), inspection for cold condensation spots on the lower surfaces immediately after the aircraft lands, ultrasonic testing using portable instruments, etc.

7. Repairability

At the aircraft design stage, especially for polymer CM structures, it is necessary to develop approaches for their repair in operation, taking into account the maintenance staff qualifications, as well as taking into account the engineering capabilities of the minimally equipped repair unit at airports, etc.

7.1.5 Polymer composite material aircraft with a safety high level and weight efficiency creation approach

The polymer CM aircraft weight and economic efficiency with the required safety and reliability level depends on the level of permissible maximum operational stresses σ_{max}^o (strains ε_{max}^o) that determines the structural weight, as well as on the costs S_{insp} for detecting manufactural defects and operational damage, which are included along with other costs in the cost of its production and operation (Figure 7.4).

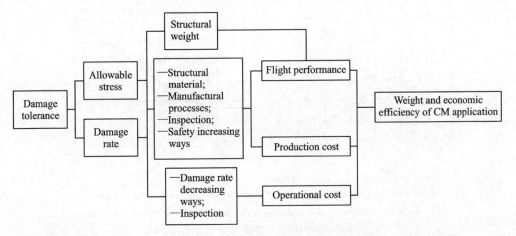

Figure 7.4   General scheme for polymer CM application effectiveness increasing

The σ_{max}^{o}, S_{insp} are monotonic functions of the polymer CM aircraft damage rate during its production and operation, which is the damage tolerance size $2L_{d.tol}$. The σ_{max}^{o}, S_{insp} are decreasing with the growth of the $2L_{d.tol}$. The σ_{max}^{o} decreasing leads to the weight efficiency decreasing. The S_{insp} increasing leads to the detecting defects (damage) cost decreasing, as a consequence, increases the structure economic efficiency. Therefore, for a required safety level, there are optimal values of $(\sigma_{max}^{o})_{opt}$, $(2L_{d.tol})_{opt}$, when the weight efficiency requirements increasing is impractical due to the cost significant increasing, and vice versa.

The optimal values of $(\sigma_{max}^{o})_{opt}$ and $(2L_{d.tol})_{opt}$ are determined primarily by the allowable stresses level, inspections laboriousness during production and in operation on the value of $2L_{d.tol}$.

Therefore, as shown in Figure 7.5, the main directions for the weight and economic efficiency increasing of the polymer CM using:

1) the design and technological approaches for the damage tolerance increasing of damaged aircraft polymer CM structures;

2) the inspection efficiency increasing over the aircraft polymer CM structures integrity during their production and operation, reducing its implementation cost.

These approaches implementation is directly related with the "operational survivability" principle application into the design process. The operational survivability specifics are parameters that establish the relationship between the design and manufactural approaches effectiveness with economic efficiency. They also determine the loading increasing feasibility of aircraft polymer CM structures, since it is limited due to increased production and maintenance costs. The dependencies of the weight, the aircraft polymer CM structures cost, its inspections and repairs laboriousness changes on the relative allowable stresses are

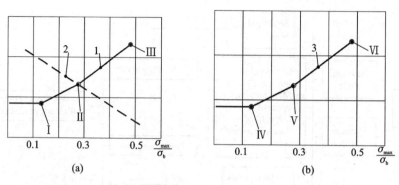

(a)　　　　　　　　　　　　　　(b)

I —epoxide polymer CM; II —crack arrest strip, needle joints, nondestructive inspection;
III —advanced CM; IV —visual inspection and repair on the mechanical joints base;
V —visual inspection and repair on the glue joints base;
VI —tooling inspection and repair on the glue joints base

Figure 7.5　Dependence of the changes in the cost (1), weight (2) and labor intensity (3) of the aircraft polymer CM structures maintenance on the allowable relative stresses increasing

shown in Figure 7.5. The required structural safety level in the form of the service life conditions, as well as the allowable costs for inspections, repairs, modifications in the production and operation processes are limitations for solving this optimization task.

The design and technological support of the aircraft polymer CM structure operational survivability can be formulated as follows: achieving the maximum weight and economic efficiency of the polymer CM using based on conditions determining and ensuring the minimum weight critical structures operation in the presence of defects and damage during the required service life in given conditions while fulfilling restrictions on safety, reliability and maintainability.

The given task solution as applied to aircraft polymer CM structures has a features number that related with the damages and destructions of the polymer CM.

The most important factors affecting to the aircraft polymer CM structures operational survivability are:

1) manufactural (materials, manufactural processes, quality inspection);

2) design methods and criteria;

3) operational (external loads, mechanical impacts, environment, routine inspections and repairs).

The main part that unites these factors is design and technological solutions that determine:

1) damageability of the polymer CM during production and operation;

2) static strength and service life indicators;

3) structural manufacturability, including the quality inspection efficiency;

4) maintainability (inspection and servicing).

The polymer CM application experience shows that the tasks of ensuring their static strength and service life should be solved simultaneously by the maintainability appropriate properties selection, since:

1) due to the polymer CM high sensitivity to stress concentrators, the required operational survivability level cannot be achieved only by applying the static strength safety factor, which obtained taking into account the increased coefficient of CM properties variation ($f = 1.5f_{add}$);

2) to ensure the polymer CM structures survivability, that designed without taking into account damage rate during production and operation, such approaches as partial adjusting or service life limiting are either unacceptable or ineffective.

The design and technological support of task solution for the aircraft polymer CM structures operational survivability can be represented as the implementation of separate procedures:

1) effect determination of the manufactural and operational factors, the design approaches and criteria on the polymer CM damageability;

2) operational survivability determination for the given polymer CM damageability conditions, taking into account the design and technological factors effect;

3) ensuring the required operational survivability for the aircraft polymer CM structures of the minimum weight with a given safety level and while meeting the inspections and repairs labor-intensiveness limitations during production and in operation.

This can be done only by specific approaches applying at the preliminary design stage, during the manufacture and during its operation:

1) deep analysis on the actual damageability conditions of polymer CM structures during production and operation;

2) obtaining real damageability and survivability features of polymer CM structures;

3) structure certification and safety inspection procedure optimization in operation.

Thus, the following main elements should be developed:

1) requirements for the survivability, safety and effectiveness criteria of the polymer CM structure application;

2) a probabilistic-statistical model of polymer CM structure damage and survivability under production and operating conditions;

3) a system for ensuring the polymer CM structures strength and service life, including two parts: synthesis and monitoring;

4) the results of actual production and operation analysis;

5) feedback link of the polymer CM structures production and operation experience with all elements of the problem.

7.2   Testing and certification of the polymer composite material

Polymeric materials and parts made from them are used as goods and must have high quality and be safe for human using (polymer packaging for food products, children's toys, dishes, jewelry, clothing, computers, electrical products, building materials, craft parts, etc.).

The certification (include polymer CM) is related with the guarantees provision to the consumer that the goods (parts) that he purchased meet the specific standards requirements.

Products and services certification have become mandatory. It is considered as an official quality confirmation, and safety. It largely determines the products competitiveness, and hence profitability, efficiency and development of the production.

Certification is a conformation procedure of the product, production or service compliance with regulatory requirements, by means of which a third party officially proves that the product, production (process) or service meets the specified requirements.

Third party in the certification procedure means an independent, competent organization that assesses the products quality for the participants in the "sale-purchase" process. The first party is considered to be the products manufacturer or seller; the second is the buyer, the consumer.

Certification is considered the main reliable way of the product (process, service) conformity proving to the specified requirements. The procedures, rules, tests and other procedures that can be considered as the certification process part itself may vary depending on a factors number. Among them: legislation related to standardization, quality and certification itself; certification object features, which in turn determines the test method choice, etc. In other words, conformity proof is carried out according to one or another certification system.

The certification system (in general) consists of:

1) the organization that manages the system, oversees its operations and can

transfer the right to conduct certification to other organization;

    2) rules and procedure for certification;

    3) normative documents for compliance with which certification is carried out;

    4) certification procedures;

    5) the inspection procedures.

The conformity assessment is the compliance degree systematic check with specified requirements. Another concept of the conformity assessment is checking, which is considered as a conformity assessment by measuring of the product specific properties.

The most reliable test results are "third party" tests for conformity assessment.

Testing is a technical operation (procedure), which consists of one or more properties given product (process or service) determining in accordance with an established procedure according to accepted rules. The tests are carried out in special testing laboratories.

7.2.1   Composite material behavior

1. Mechanical behavior

In detail, they were considered in Chapters 1, 2 and 4. Here will be explanation of them with taking into account their testing.

Mechanical properties determine the physical body behavior under the applied load action. Numerically, this behavior is assessed by strength and deformability. Strength characterizes resistance to destruction, and deformability characterizes the change in the CM sample size caused by a load applied to it. Both strength and deformation are a function of one independent variable-external load.

The elasticity modulus ($E = \sigma/\varepsilon$) is an integral characteristic that gives a concept of the structural material rigidity. The single elasticity modulus does not give a complete picture of CM capabilities, since as ε tends to zero, the value of E can reach large values even for materials with low strength. This effect is observed, for example, for brittle materials (ceramics, cast iron, plastics at temperatures below the frost resistance temperature). Typical tensile curves are shown in Figure 7. 6.

Analyzing curve 1, it can be seen that the polymer CM during loading first shows elastic properties (part $O—A$), when the stress is proportional to the specific elongation, that is, Hooke's equation is provided ($E = \sigma/\varepsilon$). With a further elongation increasing in the polymer CM, in addition to the elastic one, a plastic component appears, that causing graph bending (part $A—B$, curve 1). The humped shape of graph (part $B — C$), as a rule, is caused by the "necking down" formation, which is observed for the high density polyethylene, polypropylene or polylactam tension. At further elongation increasing, the "cold flow" (part $C— D$), is occurred at practically constant stress and is accompanied by a change in the polymer supramolecular structure. Graph transformations can also occur during further elongation, accompanied by the new "necking down" formation at the fully elongated CM sample part, with its followed tension and breaking (section $D — E —F$).

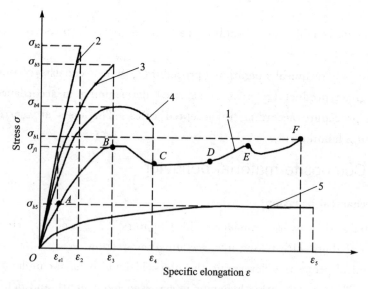

Figure 7.6　Typical tensile graphs

　　Depending on the polymer CM physics-chemical properties, its structural and compound features, the CM deformation-strength graphs can be separate parts scaled analogs of the "full curve" 1.

　　So, for example:

　　1) the behavior of plastics such as poly methyl methacrylate, phenolic plastics, aromatic polyamide is described by graph 2;

　　2) the behavior of polystyrene, cured epoxy, reinforced epoxy plastics is described by graph 3;

　　3) amorphous-crystalline low-density polyethylene, rigid polyvinyl chloride, nylon-epoxy novolac block copolymers under tension have behavior similar to graph 4;

　　4) highly plasticized polyvinyl chloride, low-density polyethylene, epoxy copolymers with elastomers have behavior similar to graph 5.

　　It should be noted that the graphs in Figure 7.6 are obtained under standard test conditions.

Impact resilience characterizes the materials ability to resist loads applied at high speed. In the practice of CM properties evaluation, the transverse impact test, which is carried out at the rebound pendulum machine, has found the widest application. In this case, samples in the form of the standard notched bar (Charpy impact strength) or the unnotched sample are tested on a standard rebound pendulum machine, which has two supports for sample mounting. The impact is applied by the pendulum in the sample middle.

Hardness characterizes the surface mechanical properties and is one of the polymer CM additional parameter. The CM effective application possible ways are evaluated by hardness. Plastics that are soft, elastic, with low hardness, are used as sealing, tight rope and gasket materials. Hard and strong plastics can be used for the structural parts production, such as gear wheels and rims, heavy-duty bearings, threaded joints components, etc.

The main mechanical properties of various polymeric materials are given in Tables 7.1—7.4

Table 7.1　Mechanical behaviors of general purpose thermoplastics

Material grade	Tensile break stress/ MPa	Tensile elasticity modulus/ GPa	Bending elasticity modulus/ GPa	Notch toughness/ $(kJ \cdot m^{-2})$	Unnotched toughness/ $(kJ \cdot m^{-2})$	Breaking elongation/ %	Ball hardness/ MPa
Low-density polyethylene	7—17	0.09—0.01	0.09—0.13	without damage	without damage	50—600	14—23
High-density polyethylene	13—30	0.55—0.65	0.65—0.75	without damage	without damage	5—1,000	44—52
Polypropylene	24—39	0.8—1.18	1.2—1.7	3.5—80	30—80	10—800	40—70
Polyvinyl chloride	40—120	2.6—3.0	—	2—10	40—80	5—100	10—160
General-purpose polystyrene	30—48	—	2.7	1.4—2.0	17—28	1.5—4	140—160
Styrene acrylonitrile	50—85	3	3	14—23	16—24	3—5	16—24
Styrene methylstyrene	—	—	2	—	—	2	—

(continued)

Material grade	Tensile break stress/ MPa	Tensile elasticity modulus/ GPa	Bending elasticity modulus/ GPa	Notch toughness/ $(kJ \cdot m^{-2})$	Unnotched toughness/ $(kJ \cdot m^{-2})$	Breaking elongation/ %	Ball hardness/ MPa
Polystyrene-co-butadiene acrylonitrile	32—65	1.5—2.6	1.5—2.8	5—25	8—100	12—70	10—200
Polyvinyl acetate	20—50	1.3—2.3	1.3—2.3	—	5—8	10—20	20—50
Polyvinyl butiral	22—60	—	2.0—2.2	—	80—125	15—180	100—110
Polyvinyl fluoride	60—70	—	3.3—4.0	—	15—30	5—11	190
Polyviol	60—140	—	5.5	—	4—6	5—6	—
Poly methyl methacrylate	55—115	2.7—4.2	—	1.6—1.8	2—30	1.5—23	10—300

Table 7.2　Mechanical behaviors of engineering thermoplastics

Material grade	Tensile break stress/ MPa	Tensile elasticity modulus/ GPa	Bending elasticity modulus/ GPa	Notch toughness/ $(kJ \cdot m^{-2})$	Unnotched toughness/ $(kJ \cdot m^{-2})$	Breaking elongation/ %	Ball hardness/ MPa
PA6 (Russia) Grilon CA6EH (USA)	50—65	1.6—2.4	1.5—2.3	5—10	without damage	150—300	100—110
PA6NS (Russia) Maranyl AD385 glass-filled (GB)	120—150	5.0—8.5	5.0—8.0	5—10	30—50	2—7	130—150
PA610 (Russia) Maranyl A castable (GB)	50—60	1.5—1.7	1.5—1.7	5—10	without damage	100—200	110—150

(continued)

Material grade	Tensile break stress/ MPa	Tensile elasticity modulus/ GPa	Bending elasticity modulus/ GPa	Notch toughness/ (kJ · m^{-2})	Unnotched toughness/ (kJ · m^{-2})	Breaking elongation/ %	Ball hardness/ MPa
PA12 (Russia) Grilamid TR (USA)	45—55	1.2—1.6	1.2—1.4	5—20	without damage	200—260	80—105
Polycarbonate	60—65	2.4	—	20—30	without damage	80—120	100—110
Polyethylene- terephthalate	50—70	2.8—3.0	2.8—3.0	2—3	22—90	5—50	115—130
Polybutylene terephthalate	55—60	2.2—2.7	2.0—2.3	2.0—4.5	100—120	30—120	100—110

Table 7.3　Mechanical behaviors of engineering thermoplastics heat—resistant polymer

Material grade	Type	Tensile break stress/ MPa	Tensile bending stress/ MPa	Tensile elasticity modulus/ GPa	Bending elasticity modulus/ GPa	Notched Izod impact resistance/ (J · m^{-1})	Breaking elongation/ %
Polysulfones	base	70	106—109	2.5	2.7	69	75
	Glass-filled 30%	101—180	140—154	7.4	7.6	59—75	1.5—3.0
Polyetherimide	base	105	118—171	3	2.7—3.3	50—53	60—80
	Glass-filled 30%	165	230—237	9	8.3—9.0	100—107	3
Polyethersu-lfone	base	84—86	129	2.4—2.5	2.6	—	50—60
	Glass-filled 30%	140—145	190—192	—	2.4	—	3
Polyimide sulfone	base	70	—	4.8	—	—	—
Polyarylethe-retherketone	base	80—90	—	3.2	4	40	100
	Glass-filled 30%	166—176	—	8.6	9—10	98	3
Polyphenylene sulfide	base	76	—	2.2—2.7	3.8—4.2	25—70	1—3
	Glass-filled 30%	66—135	—	5.3—7.9	11.7—12.6	60—80	0.9—4

Table 7.4　Mechanical behaviors of thermosetting CM material

Material grade	Tensile break stress/ MPa	Tensile elasticity modulus/ GPa	Bending elasticity modulus/ GPa	Notch toughness/ $(kJ \cdot m^{-2})$	Unnotched toughness/ $(kJ \cdot m^{-2})$	Breaking elongation/ %	Ball hardness/ MPa
Phenoplasts	22—127	6—12	6—17	1.7—14	3—21	0.1—7.0	200—650
Aminoplastics	30—80	7.5—10	10—13	—	6—7	0.2—0.6	200—450
Polyeurethanes	1.2—56	—	—	—	5—20	240—600	—
Silicone plastic	18—30	—	—	—	2—80	2—80	100—200
Silicone plastic glass-filled	10—70	—	—	—	—	—	—
Cured epoxy plastics	14—90	—	—	—	4—40	1—4	100—200
Epoxide compound	6—67	0.3—300	1—2	—	3—18	0.3—70	20—200
Fabric-reinforced laminate	34—130	4.0—6.5	3.5—6.4	15—40	10—50	1	200—350
Asbestos-reinforced laminate	42—250	4—20	3.7—20	—	8—34	—	190—450
Glass-cloth-base laminate	—	9.7—31.0	—	—	—	1.0—1.5	—
Glass-fiber plastic	30—670	1.4—3.5	18—25	—	12—250	1.5	200—450
Cured carbamide resins	40—56	—	5.0—5.5	—	2.5—6.0	0.5	—
Cured polyester resins	6—70	1.7—4	—	—	—	0.5—80	30—250

2. Thermal behavior

Often, the polymer CM thermal behavior is more important than strength behavior. The polymer CM part operational temperature affects the breaking stresses, deformability, elasticity modulus, hardness, impact strength and other behaviors.

1) If the thermoplastic physics-mechanical behaviors are lower, then they are more sensitive to temperature changes. Thus, among polyolefins, polypropylene, which is a structural material with accordance to its strength and stiffness, loses about 25% of its bending strength when it is heated up to 80 ℃. The high-density polyethylene loses about 50% of its original strength even at 60 ℃. Similar ratios are observed in tensile and bending tests of polyolefins.

2) Amorphous polymers, in general, are shown with a slightly lower dependence of the strain-strength behaviors on temperature. Plastics such as polycarbonate, polyethylene terephthalate, polysulfone are kept more than 70% strength at temperatures over 100 ℃.

3) The chopped fiberglass applying into thermoplastics (content up to 30%) helps to reduce the behavior heat dependence, and not only for heating, but also at negative temperatures up to − 60 ℃. And, in this case, the polymer binder behaviors determine the CM behavior.

4) Temperature has a very significant effect on the thermoplastics rigidity that depended on their elasticity modulus. The crystalline thermoplastics deformability [such as high-density polyethylene, PA12, PA66 (Russia)] changes sharply even with relatively small temperature variability in a range (− 40—+ 40 ℃). This specific must be taken into account when choosing polymer CM for the structural parts manufacturing, especially that operate under long-term cyclic stresses [PA66 (Russia), polycarbonates].

5) The thermoplastics impact toughness dependence on temperature is opposite to the elastic modulus behavior, that is, it increases with increasing, and it decreases with the temperature decreasing.

The temperature effect on thermosetting plastics is primarily depended on the binder ℃. It is known that in the vitreous state the deformation and strength behaviors slowly decrease with the temperature increasing. The softening temperature exceeding is affected on anaccelerated behavior decreasing. In some cases, in the temperature range of 10—20 ℃, the elasticity modulus and breaking stress are decreased by two-order reduction. The dispersed fibers, which have an additive effect on the CM, smooth this sharp decreasing. In highly filled reinforced plastics, the transition of the binder from the vitreous state to a highly elastic state occurs even more slowly.

Polymer CM temperature behaviors are given in Tables 7.5—7.8.

Table 7.5 Temperature behaviors of general-purpose thermoplastics

Material grade	Upper operational temperature limit/℃	Lower operational temperature limit/℃	Vicat softening temperature/℃	Martens yield temperature/℃
Low-density polyethylene	60—70	−120—−45	80—90	—
High-density polyethylene	70—80	−150—−60	15—128	—
Polypropylene	95—110	−50—−5	—	—
Polyvinyl chloride	60—85	−20—−10	70—85	65—70
General-purpose polystyrene	65—70	−40	82—105	70—80
Acrylonitrile butadiene-styrene resin	75—85	−60	—	78—95
Polymethyl methacrylate	60—130	−50—−80	90—130	90—95

Table 7.6 Temperature behaviors of engineering thermoplastics

Material grade	Upper operational temperature limit/℃	Lower operational temperature limit/℃	Bending temperature under loading ×1.82/(MPa·℃$^{-1}$)	Vicat softening temperature/℃	Martens yield temperature/℃
PA6 (Russia) Grilon CA6EH (USA)	80—105	−20	45—70	170—200	75—76
PA610 (Russia) Maranyl A castable (GB)	80—120	−40	65—70	170—210	55—60
PA12 (Russia) Grilamid TR (USA)	70—80	−60	55	170	—
Polycarbonate	150—135	−120	104—109	145—150	120—145
Polybutylene terephthalate	75—120	−60	50—55	190—195	—

Table 7.7   Temperature behaviors of engineering thermoplastics heat-resistant polymer

Material grade	Type	Long-term operation temperature/℃	Bending temperature under loading $\times 1.82/(MPa \cdot ℃^{-1})$	Softening temperature/℃
Polysulfones	base	140—150	172	—
	glass-filled 30%	140—150	175—181	—
Polyimides	base	220—265	360	—
	glass-filled 40%	250—265	Higher 300	—
Polyarylates	base	160	180—355	190
Polyetherimides	base	170—180	198—200	217
	glass-filled 30%	175	208—210	—
Polyester-sulfones	base	180	201—203	—
	glass-filled 30%	180—190	214—216	—
Polyimide sulfone	base	—	179—199	—
Polyester-ketones	base	260—300	186	160
	glass-filled 30%	260	358	—
Polyaryletheretherketone	base	230—290	160—315	143—200
	glass-filled 30%	240—290	286	—
Polyphenylene sulfide	base	185	135—138	185
	glass-filled 40%	170—210	251—262	—
Polyamide imide	base	250	275	290
Polyarylenesulphone	base	200	204—220	220
Aromatic polyester	base	200—250	230	—

Table 7.8   Temperature behaviors of engineering thermoplastics and their compositions

Material grade	Upper operational temperature limit/℃	Lower operational temperature limit/℃	Martens yield temperature/℃
Phenoplasts	66—220	−60— −40	125—250
Aminoplastics	80—130	180— −40	95—200
Polyeurethanes	90—120	−60	—
Silicone plastic	250—300	—	200—300
Silicone plastic glass-filled	300—400	—	250—320
Cured epoxy plastics	—	—	80—250
Epoxide compound	60—220	—	25—200
Fabric-reinforced laminate	105—140	−40	130—140

(continued)

Material grade	Upper operational temperature limit/℃	Lower operational temperature limit/℃	Martens yield temperature/℃
Asbestos-reinforced laminate	125—130	−40	—
Glass-cloth-base laminate	130—250	−60	—
Cured carbamide resins	—	—	160—240
Cured polyester resins	—	−40—0	—
Furane resins	300	—	—

3. Thermal-physical properties

Thermal-physical properties are extremely important for determining the polymer CM practical effect. Such polymer CM devices parts as gear wheels and pinions, plain bearing liners, friction braking systems, sealing structures and many others that operated in non-stationary thermal fields are required the thermal-physical properties polymer CM data. This is necessary to select the manufactural processes types: base substance heating or cooling (melting, currying, softening, etc.). Such properties are conventionally divided into two groups.

1) They are determined by the polymer CM external behavior when the temperature changes, for example, thermal expansion or dilatometry properties.

2) They are explained by the internal material reaction to thermal effects. The intensity of each reaction type is determined by the corresponding thermal-physical coefficient.

Thermal expansion factor explains the overall size change as a function of temperature.

The thermal conductivity factor is numerically equal to the heat amount transferred through an isothermal surface unit per time unit with the unity temperature gradient. It characterizes the material thermal inertial properties. The larger its value is, the faster the temperature is equalized at all part points. Accordingly, the lower its value is, the better the material thermal-insulating properties are. The thermal conductivity factor is necessary for manufactural purposes to assess the part cooling time or to assess the polymer CM part behavior at non-standard thermal fields.

Thermal-physical properties of some polymer CM are given in Table 7.9.

Table 7.9 Thermal-physical properties of some polymer CM

Material grade	Thermal conductivity/ $[W \cdot (m \cdot K)^{-1}]$	Heating capacity/ $[kJ \cdot (kg \cdot K)^{-1}]$	Temperature conductivity$\times$ $10^{-7}/(m^2 \cdot s^{-1})$	Average line expansion coefficient$\times$ $10^5/K^{-1}$
Low-density polyethylene	0.32—0.36	1.8—2.5	1.3—1.5	21—55
High-density polyethylene	0.42—0.44	2.9—3.1	1.9	17—55
Polypropylene	0.19—0.21	1.93	1.3	11—18
General-purpose polystyrene	0.09—0.14	1.16—1.3	0.94	6—7
Acrylonitrile butadiene-styrene resin	0.12	1.24	0.9	8—10
Polyvinyl chloride	0.16	1.11	0.118	6—8
Polypyrrolidone	0.38	2.0	1.73	12—30
Polyethylene terephthalate	0.2	0.99	1.56	8—13
Polymethyl methacrylate	0.19—0.2	1.3—2.1	0.9—1.1	7—12
Polycarbonate	0.31	1.37	0.8—1.9	2—6
Phenolic plastics	0.2—0.5	1.0—2.3	0.9	1.0—4.0
Amino plastics	0.28—0.34	1.1	0.95	1.5—3.3
Epoxy plastics	0.3—0.42	—	—	0.8—2.5

4. Chemical fastness

The list of aggressive agents affecting the polymer CM behavior is wide. They can be divided into some groups. These are: mineral and organic acids; the water organic acids solutions; alkalis and oxidizing agent's solutions; aliphatic and aromatic solvents; fuels and lubricants. The aggressive agents effect on a polymer CM can be accompanied by its expanded, diffusion into the polymer CM and chemical interaction leading to its destruction.

There are standards for the chemical resistance determining of a polymer CM. The higher score is, then the chemical material's resistance is higher too. Usually polymer CM resistance to aggressive agents is assessed by their mass changing on a five-point scale base: 5 —high resistance; 4 —satisfactory; 3 —the material is not stable in all cases; 2 —durability is insufficient, not recommended for use; 1 —the material is not resistant and quickly destructors.

Fluor plastics have high chemical inertness and resistance to destruction. Polyolefins such as low-density polyethylene, high-density polyethylene, polypropylene and unplasticized polyvinyl chloride, also have significant chemical resistance. Polycarbonate and polystyrene plastics are somewhat inferior to them. Heterogeneous chain polymers such as polyamides are destroyed by hydrolytic degradation and active expansion due to their hydrophilicity. Structural thermoplastic, formaldehyde polymer, cannot be resistant to aggressive agents. Thermosetting plastics are sensitive to alkaline and oxidizing solutions. At the same time, antegmites and faolites that are high-filled with powdered graphite, which are based on phenol formaldehyde or phenolaldehyde binder, are widely used in chemical industry.

Reinforced polymer CM can be used for a long time in acids and alkali solutions with a concentration of up to 10%, as well as in solvents and fuels or lubricants.

5. Electrical properties

Electrical properties are parameters' set that characterize the polymer CM behavior in electromagnetic field. The following parameters are often used: dielectric capacitivity, dielectric loss, dielectric loss factor, electrical conductivity and electrical strength, and tracking resistance.

The dielectric capacitivity ε is equal to the ratio of the electric capacitor capacitance, between the plates of which is a polymer CM, to the same capacitor capacitance, between the plates of which vacuum or air. By the ε, all polymeric materials are conventionally divided into groups:

1) non-polar, $1.8 < \varepsilon < 2.3$;
2) low-polar, $2.3 < \varepsilon < 3.0$;
3) polar, $3.0 < \varepsilon < 4.0$;
4) high-polar, $\varepsilon > 4.0$.

The division conventionality and the polymer CM electrical properties strongly depend on external conditions —temperature, humidity, the environment ionization rate, electric field intensity, current intensity, etc. For the standardized measurements, the electromagnetic field frequency is 10 Hz; the temperature is 20 ℃; the relative humidity is 60%.

6. Polymer flammability

The main indicators that determine the polymer flammability are as follows:
1) flammability index (K);

2) firing point (t_f);

3) self-ignition point (t_{s-i});

4) oxygen index (OI).

In addition, from the fire hazard, important indicators are the smoke emission and gaseous release during polymer firing.

All organic substances are flammable, in order to reduce the fire hazard of polymer CM; they are modified: various additives, such as flameproofing initiators, are introduced that affect the polymers firing or inhibit it. Methods for assessing the polymers flammability are determined by standards —10456, 17088, 21207, 21793 (Russia); ASTMD 2863 (USA); DJN 22117 (Germany); AFNOR NFT 51071 (France), etc.

The flammability index is a dimensionless value that is the ratio of the heat released during firing amount to the heat spent on material sample igniting amount. Materials with a K greater than 0.5 are flammable materials.

The firing point and *self-ignition point* characterize the polymer CM behavior with and without fire.

Polymers firing is accompanied by significant *smoke emission*. The material smoke emission is determined by the specific absorbance, which shows the illumination attenuation at a distance of 1 m from the light source in the smoke emitted during the firing of 1 kg of material in a room with volume 1 m^3.

The oxygen index of polymers is widely used to assess their flammability. It shows at what minimum amount of oxygen in the O_2-N_2 mixture a vertically located sample ignites when it is ignited. The oxygen index is expressed as a percentage and depends on the polymer chemical structure and its content in the material. Polymers with OI less than 27 are highly firing, but if OI is less than 20, then firing proceeds is high speed, but if OI equals 20—26 it is slowly. Polymers with an OI of more than 27 are self-extinguishing when taken out of the fire and are considered to be hardly firing.

7. Tribological properties

Tribotechnical properties characterize the polymer CM applicability in friction units. The friction processes and accompanying wear are numerically estimated by the following parameters:

1) Friction ratio (sliding) m. Numerically $m = \dfrac{F}{N}$, where F is the friction force, and N is the normal force. By its value, all triboplasts are subdivided into frictional materials (m more than 0.3) and antifrictional materials (m less than 0.2).

2) The wear rate J is equal to the ratio of the wearing part size or mass change or energy consumption for this process to the friction path L, also called runin. There are the wear rates: linear, mass, energy.

3) Wear life is the inverse value of the wear rate. There are: linear wear life (dimensionless), mass wear life (m/mg) or energy wear life ($\frac{m}{J}$).

4) The friction mode parameters: nominal pressure $P = \frac{N}{S}$, Pa; sliding speed V, m·s^{-1}. The product ($P \cdot V$) is very important in practical applicability assessing of the material in plain bearings, that operated under certain energy conditions of load P and sliding speed V. ($P \cdot V$) has units W·m^{-2}. Each triboplast is characterized by its own allowable mode parameter of ($P \cdot V$).

5) Friction conditions are determined by the lubricating layer presence on the friction surface. There are:

① dry friction —the lubricating layer is completely absent (implemented in vacuum);

② boundary friction —on the sliding surface there is a layer of moisture condensed from air (or other lubricant) of molecular thickness;

③ half-floating condition —surfaces are partially separated by a lubricating layer;

④ fluid friction —the surfaces are completely separated by a lubricating layer, the thickness of which exceeds the height of surfaces microirregularity and roughness.

Friction conditions significantly affect the friction ratio. The friction ratio can be reduced in many times even by a boundary lubricant layer in the friction zone.

7.2.2 Composite material test methods

1. Mechanical tests

1) *Strength, deformation and elasticity modulus at tension.*

They are based on standards: ISO R527 (DIN 53455, DIN 53457, ASTM D638M).

Tensile stress and strain relationships are the most widely published mechanical properties for comparing materials or designing specific parts.

Test speeds:

① Speed A —1 mm/min —tensile elasticity modulus.

② Speed B —5 mm/min —tensile stress diagram for glass-fiber filled resins.

③ Speed C —50 mm/min — tensile stress diagram for resins without reinforcement.

Tensile stress-strain relationships are determined as follows. The dumb-bell sample (Figure 7.7), is tinseled with a constant rate and the applied load and elongation are recorded.

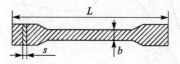

$L = 150$ mm; $b = 10$ mm; $s = 4$ mm

Figure 7.7   Test sample ISO R527

After that, the stresses and strains are calculated:

① Stress: load divided to unit of area of the initial cross-section, MPa (Figures 7.8 and 7.9);

② Strain: (Elongation divided to original length) × 100%.

Other mechanical behavior, determined from the stress-strain dependence, are:

① Modulus: Stress divided to strain, MPa;

② Yield strength: Maximum stress of the beginning of the material plastic flow, MPa;

③ Tensile strength: stress at destruction, MPa;

④ Destructive strain: Destruction strain or maximum elongation, %;

⑤ Proportional elastic limit: The point where nonlinearity begins, MPa;

⑥ Elasticity modulus: Modulus below the proportional limit, MPa.

Figure 7.8   Testing equipment for the tensile test

2) *Strength and elasticity modulus at bending.*

They are based on the standards: ISO 178 (DIN 53452, ASTM D790).

Flexural strength can show how well a material resists bending, or "how stiff

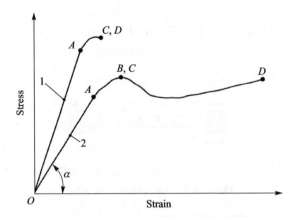

1—sample from polymer CM with glassfiber reinforcement (brittle break);
2—sample from polymer without reinforcement (elastic break);
A—proportional elastic limit; B—yield strength; C—tensile strength; D—break point;
zone O—A flexural yield stress zone, elastic behavior; tan α—elasticity modulus

Figure 7.9　The stress-strain diagram

the material is". Unlike tensile loading, in flexure tests, all loads act in one direction. The sample is free-ended rod which is loaded in the middle of the span: this creates a three-point loading configuration. On a test equipment, the loading tip presses the sample with a constant speed of 2 mm · min^{-1} (Figure 7.10).

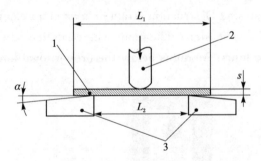

1—sample; 2—tool; 3—base
$L_1 = 80$ mm; $L_2 = 64$ mm; $s = 4$ mm; $α = 5°$

Figure 7.10　Flexural test layout

To calculate the bending elasticity modulus, a strain versus load curve is plotted from the recorded data. It is used a minimum of five load and strain values starting from the curve original linear part.

Flexural modulus (the ratio of stress to strain) is most often used for elastic properties. The flexural modulus is equivalent to the slope of a line tangent to the stress-strain diagram in this curve part where the sample has not yet deformed (Figure 7.11).

The unit of stress and flexural modulus is MPa.

Figure 7.11　Flexural testing equipment

3) *Wearing test with the Taber equipment.*

It is based on the standards: ISO 3537 (DIN 52347, ASTM D1044).

The abrasion loss is measured by sample abrasion with the Taber machine (Figure 7.12).

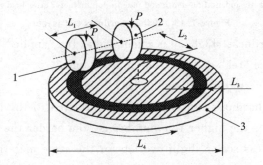

1—abrasive wheel CS 17 Calibrise turned clockwise; 2—abrasive wheel CS 17 Calibrise turned anticlockwise;

3—sample

$P = 5$ N; $L_1 = 80$ mm; $L_2 = 19$ mm; $L_3 = 10$ —12 mm; $L_4 = 120$ mm

Figure 7.12　Wearing test with the Taber equipment layout

The sample is fixed on a disk turning at 60 rpm. The loads are generated by the special weights press of the abrasive wheels to the sample.

After the specified number of cycles, the test is terminated. Abrasion loss is defined as the particles' mass that has been removed from the sample. It has mg · $(1,000 \text{ cycles})^{-1}$ unit.

Abrasive wheels are actually round-shaped grindstones. Various circles types are used.

4) *Hardness testing.*

There are Brinell hardness, Rockwell hardness, Shore hardness. The Rockwell test determines the polymers hardness after strain elastic recovery. This is the difference between this way and the Brinell and Shore hardness tests: in these tests, the hardness is determined by the strain depth under load and, therefore, excludes any strain elastic recovery of the material. Therefore, Rockwell numbers cannot be directly correlated with Brinell or Shore hardness numbers.

Shore's A and D hardness numbers ranges can be compared with Brinell's hardness numbers ranges. However, there is no linear correlation.

Brinell hardness test is based on ISO 2039-1 (DIN 53456).

A polished hardened steel ball that is 5 mm in diameter is pressed into the test sample surface (at least 4 mm thick) with a force of 358 N. The ball imprint depth is measured 30s after the load application (Figure 7. 13).

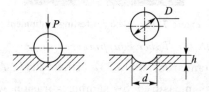

P —applied load; D —ball diameter;

h and d are indentation depth and indentation diameter after load removal

Figure 7. 13   Brinell hardness test layout

The Brinell hardness H358/30 is calculated as the "applied load" divided by the "ball imprint surface area" and has $(N \cdot mm^{-2})$ unit.

Rockwell hardness test is based on ISO 2039-2.

The Rockwell hardness number is directly related to the ball imprint size on the polymer surface: the higher this number is, the harder the material is. Due to the Rockwell hardness scales slight overlap for the same material, it is possible to receive two different numbers on two different scales, and both of these numbers may be correct.

The indenter, which is a polished hardened steel ball, is pressed into the test sample surface (Figure 7. 14). The ball diameter depends on the Rockwell scale used. The sample is loaded with "light load", then with "base load", then again with the same "light load".

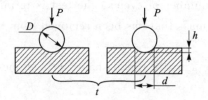

P—applied load; D—ball diameter;

h and d are indentation depth and indentation diameter; t = 30 s without elastic recovery

Figure 7. 14   Rockwell hardness test layout

The actual measurement (Figures 7. 15) is based on the ball imprint total depth, which is calculated as the total depth after the base load is removed minus the elastic recovery after the base load is removed and minus the ball imprint depth at light load. The Rockwell hardness number is calculated as "130 minus the ball imprint depth in units of 0.002 mm".

(a) portable Rockwell hardness tester　　　　　　　(b) laboratory Rockwell hardness tester

Figure 7.15　Rockwell test equipment

Rockwell hardness numbers should be between 50 and 115. Values outside this range are considered inaccurate: the measurement must be repeated again using the next harder scale. The scales increase in hardness from R through L to M (with increasing material hardness). The loads and diameters of the indenters are shown in more detail in Table 7.10.

Table 7.10　The intender load and diameter

Hardness scale	Light load/N	Base load/N	Rockwell ball diameter/mm
R	98.07	588.4	12.7
L	98.07	588.4	6.35
M	98.07	980.7	6.35

If a softer material requires a less hard scale than the R scale, then the Rockwell hardness test is not suitable. The Shore hardness test method (ISO 868) can then be used, which is used for low modulus materials.

Shore hardness test is based on ISO 868 (DIN 53505, ASTM D2240).

Shore hardness numbers are dial indications obtained by the certain steel rod penetration into the polymer. This hardness is determined by scleroscopes of two types, both of which have calibrated springs to apply a load to the indenter. Scleroscope A is used for softer materials and scleroscope D is used for harder materials.

Shore hardness numbers change (Figure 7.16):

① from 10 to 90 for the type A Shore scleroscope —soft materials;

② from 20 to 90 for the type D Shore scleroscope —hard materials.

　　If the measured hardness numbers are higher than 90 A, the material is too hard and the D scleroscope must be used.

　　If the measured hardness numbers are lower than 20 D, the material is too soft and the scleroscope A must be used (Figure 7.17).

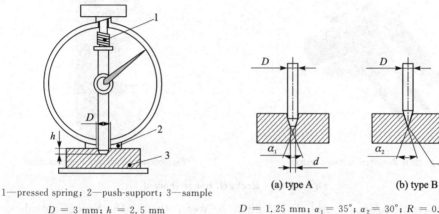

1—pressed spring; 2—push-support; 3—sample

$D = 3$ mm; $h = 2.5$ mm

Figure 7.16　Shore hardness test layout

(a) type A　　　　　　(b) type B

$D = 1.25$ mm; $\alpha_1 = 35°$; $\alpha_2 = 30°$; $R = 0.1$ mm

Figure 7.17　Scleroscope indenters

　　There is no simple relationship between the hardness measured with this test way and other basic material properties.

　　5) *Impact strength tests*.

　　In standard tests such as tensile and flexural tests, the material absorbs energy slowly. In reality, materials very often quickly absorb the applied load energy, for example, loads from falling objects, impacts, collisions, falls, etc. The impact tests purpose is to simulate such conditions.

　　Izod and Charpy ways are used to study the certain samples properties under specified impact stresses, and to assess the sample brittleness or impact strength. Test results from these tests should not be used as a data source for part's design calculations.

　　Data on typical material properties can be obtained by testing different types of test samples prepared under different conditions, such as, a notch radius and test temperature changing.

　　Tests for both ways are carried out on a power pendulum (Figure 7.18). The sample is clamped, and a pendulum with a hardened steel impact surface of a certain radius is released from a predetermined height, which causes the sample to shear from an impact load. The pendulum residual energy lifts it up. The difference between the drop height and the back height determines the energy expended to test sample break. These tests can be performed at regular temperature or at negative temperatures to determine cold brittleness. The test sample can be different notches types and sizes.

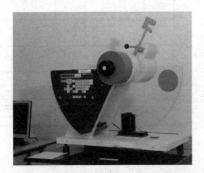

Figure 7.18   Impact strength laboratory equipment

Drop weight test results, such as the Gardner way or bent plate, depending on the geometry of the drop weight and support. They can only be used to determine the materials relative ranking. Impact test results cannot be considered absolute unless the test equipment and sample geometry meet the end-use requirements. It can be expected that the materials relative ranking for the two test methods will be the same if the destruction nature and impact velocity are the same.

Interpretation of impact test results and comparison of ISO and ASTM methods is very important process. Impact behavior can be highly dependent on sample thickness and molecular orientation. The different samples thicknesses used in the ISO (4 mm) and ASTM (3 mm)methods can have a very significant effect on the impact strength values. Thickness changing from 3 mm to 4 mm can even lead to a transition from elastic to brittle destruction behavior due to the influence of molecular weight and notched sample thickness when using the Izod way, as demonstrated for polycarbonate resins (Figure 7.19). Materials that have already shown brittle fracture at a thickness of 3 mm, for example, materials with mineral and fiberglass fillers, are not affected by the change in sample thickness. Materials with modifying additives that increase impact strength have the same properties.

It should be clearly understood that:

① not the materials have changed, but only the test methods;

② the transition from elastic to brittle destruction has an insignificant effect in reality: the overwhelming majority of designed parts have a thickness of 3 mm or less.

Izod impact test is based on ISO 180 (ASTM D256).

Izod impact tests of notched samples has become the standard way for comparing the polymers impact strength. However, this test results are little responding with the real part impact behavior in real conditions. Due to the materials' different sensitivity to notch, this test way allows some materials'

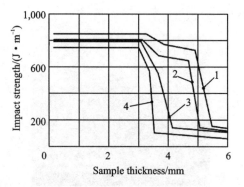

Molecular weight 1 > 2 > 3 > 4

Figure 7.19   The notched sample thickness and molecular weight influence on the impact tests results of polycarbonate resins according to Izod

rejections. While the results of these tests have often been applied as meaningful impact strength evaluation, these tests, in practice, measure the material notch sensitivity rather than the polymer withstand impact ability. These tests' results are widely used for comparing the materials impact strength. Notched Izod impact tests are best used to determine the impact strength of parts with many sharp corners, such as ribs, intersecting webs and other stress concentration points. For Izod impact tests and for unnotched samples, the same loading mode applies, except that the sample has no notch (or is clamped in an inverted position). Tests of such type always give better results than notched Izod tests due to the absence of a stress concentration.

The impact strength of notched sample according to the Izod way is the impact energy expended to the notched sample destruction divided by the initial sample cross-sectional area at the notched location (Figure 7.20). It has ($kJ \cdot m^{-2}$) unit. The sample is clamped vertically in a power pendulum.

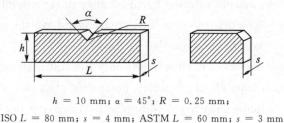

$h = 10$ mm; $\alpha = 45°$; $R = 0.25$ mm;

ISO $L = 80$ mm; $s = 4$ mm; ASTM $L = 60$ mm; $s = 3$ mm

Figure 7.20   Impact strength samples

ISO legends show sample and notch types:

① ISO 180/1A denotes sample type 1 and notch type A. As can be seen in Figure 7.19, sample type 1 has a length of 80 mm, a height of 10 mm and a thickness of 4 mm.

② ISO 180/1O denotes the same sample 1, but clamped upside down (referred to as "uncut").

Samples, used by the ASTM way, are of similar dimensions: the same notch radius and the same height, but differ in length — 63. 5 mm and, more importantly, in thickness —3. 2 mm.

ISO test results are determined as the impact energy in joules expended to break the test sample divided by the test sample cross-sectional area at the notch (Figure 7. 21). The result has $(kJ \cdot m^{-2})$ unit.

The ASTM test results are determined as the impact energy in joules divided by the notch length (i. e. sample thickness). It has $(J \cdot m^{-1})$ unit.

The practical conversion's factor is 10, i. e. 100 $J \cdot m^{-1}$ equals approximately 10 $kJ \cdot m^{-2}$.

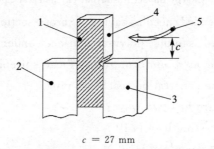

$c = 27$ mm

1—sample; 2—fixed jaw; 3—movable jaw; 4—sample impact side; 5—pendulum hammer motion

Figure 7. 21　Izod impact test layout

Charpy impact test is based on ISO 179 (ASTM D256).

The main difference between the Charpy and Izod methods is the way in which the test sample is installed. In the Charpy test, the sample is not clamped, but is placed loosely on a support in a horizontal position (Figure 7. 22).

ISO designations show sample and notch types:

① ISO 179/1C shows sample type 2 and notch type CI;

② ISO 179/2D shows sample type 2 without cut.

Samples used according to DIN 53453 are of similar dimensions.

The results for both ISO and DIN methods are determined as the impact energy in joules absorbed by the test sample divided by the test sample cross-sectional area at the notch. These results have $(kJ \cdot m^{-2})$ unit.

2. Thermal tests

1) *Vicat softening temperature.*

It is based on ISO 306 (DIN 53460, ASTM 1525).

These tests give the temperature at which the polymer begins to soften

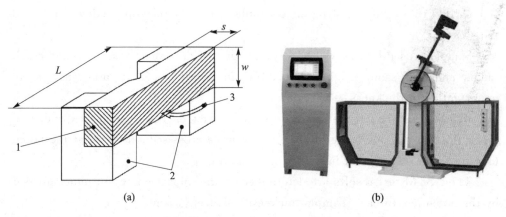

$L = 80$ mm; $s = 4$ mm, $w = 10$ mm;

1—sample; 2—supports; 3—pendulum hammer motion

Figure 7.22   Charpy impact test layout (a) and test equipment (b)

quickly. A round flat-ended needle having a cross-sectional area of 1 mm^2 is inserted into the tested sample polymer surface under a certain load. The temperature is increased at a uniform rate. Vicat softening temperature is the temperature at which the penetration reaches 1 mm.

The ISO 306 standard has two ways:

① Way A, where testing load is 10 N;

② Way B, where testing load is 50 N.

with two possible rise temperature rates:

① 50 ℃ per hour;

② 120 ℃ per hour.

ISO test results are marked as A50, A120, B50 or B120. The test sample is arranged in a heating bath with an initial temperature of 23 ℃.

After 5 minutes, a load of 10 or 50 N is applied. The bath temperatures at which the indenter tip penetrates to a depth of 1 ±0.01 mm is marked as the material Vicat softening temperature (Figures 7.23 and 7.24) at the selected load and temperature rise rate.

Some differences can be found in the ISO results compared to ASTM standards due to the different sizes of the test samples: the thermal deformation measured according to ISO ways may be lower.

2) *Thermal deformation and thermal deformation under load*

Thermal deformation is a relative measure of a material's ability to withstand a load for a short time at hot temperatures. These tests are measured by the temperature effect on stiffness: on a standard test sample, certain surface stresses are applied and the temperature is raised at a uniform rate.

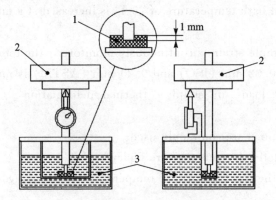

1—sample; 2—load; 3—oil bath

Figure 7. 23　Vicat softening temperature thermal test layout

Figure 7. 24　Laboratory Vicat softening temperature tester

Samples, used in tests, are annealed and unannealed. Tempering is a process in which a sample is heated to a certain temperature, it is held for some time at this temperature, and then gradually cool off to ambient temperature. Such procedures make it possible to reduce or completely remove internal stresses in the sample body, which have arisen, for example, at the accelerated polymerization process in an injection molding machine.

For both ISO and ASTM standards, the loaded test sample is arranged in a heating bath filled with silicone oil.

Sample surface stresses are:

① low, for ISO and ASTM ways are 0. 45 MPa;

② high, for the ISO way are 1. 80 MPa, and for the ASTM way are 1. 82 MPa.

Load action is allowed for 5 minutes, but this period can be omitted if the test materials do not show appreciable creep during the first 5 minutes. After 5 minutes, the initial bath temperature of 23 ℃ is increased at a uniform rate of 2 ℃ per minute.

The tested sample strain is continuously monitored: the temperature at which the strain reaches 0. 32 mm (ISO) and 0. 25 mm (ASTM) is marked as "thermal deformation under load" or simply "thermal deformation" (heat deformation temperature).

Two abbreviations are commonly used:

① DTUL is deflection temperature under load;

② HDT is the heat-distortion temperature or bending heat deformation temperature.

In general practice, the DTUL is used for ASTM results and HDT is used for ISO results.

Depending on the surface tension created, add the letters A or B to the HDT:

① HDT/A for a load of 1. 80 MPa;

② HDT/B for a load of 0. 45 MPa.

3) *Heat-distortion temperature and amorphous and semi-crystalline polymers.*

For amorphous polymers, the HDT roughly coincide with the material glass transition temperature T_g.

Since amorphous polymers have no specific melting point, they are processed in their highly elastic state at temperatures above T_g. Crystalline polymers can have low HDT and still have structural efficiency at higher temperatures. The HDT method is more reproducible with amorphous plastics than with crystalline ones. Some polymers may require the tested sample tempering (annealing) to obtain reliable results.

When glass fibers are added to a polymer, its modulus increases. Since HDT is the temperature at which a material has a specific modulus, modulus' increasing also increases the HDT. Fiberglass has greater effect on the crystalline polymers HDT compared to amorphous polymers.

Although widely used to indicate high temperature performance, HDT tests simulate only a narrow range of conditions (Figure 7. 25). In many high temperature applications, parts operate at higher temperatures, higher loads and no supports. Therefore, the results obtained with this test way do not represent the maximum application temperature, since in reality significant factors such as time, loading and nominal surface stresses may differ from the test conditions.

4) *Ball indentation HDR test (ISO 2039/1).*

These are thermal deformation tests similar to the Vicat test. The sample is arranged horizontally on a support in the heating chamber and a ball of 5 mm in diameter is pressed into it with a load of 20 N. After one hour, the ball is removed; the sample is cooled in water for 10 seconds and the imprint left by the ball is measured (Figure 7.26).

If the imprint diameter is less than 2 mm, then the material is considered to have passed the ball indentation test at that temperature.

Depending on the application, the test temperature can vary:

① 75 ℃ for non-stressed parts;

② 125 ℃ for parts under stress.

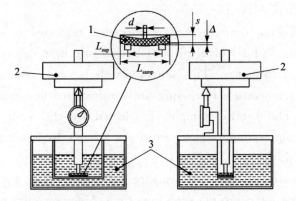

$d = 4$ mm; $s = 10$ mm; $\Delta = 0.32$ mm; $L_{sup} = 100$ mm; $L_{samp} = 120$ mm

1—sample; 2—load; 3—oil bath

Figure 7.25　HDT test layout

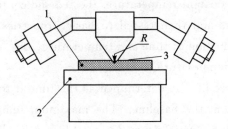

1—sample; 2—support; 3—spherical surface with $R = 2.5$ mm

Figure 7.26　Ball indentation HDR test layout

5) *Thermal conductivity (ASTM C 177)*.

Polymer thermal insulation properties are determined by measuring thermal conductivity. Wide polymer plates are arranged on both sides of a small heater. Thermal bypasses are attached to the free samples surfaces. Thermal insulators located around the test chamber prevent radial heat loss. The axial heat flux through the polymer plates can then be measured. Results are reported in W · $(m \cdot ℃)^{-1}$.

6) *Relative temperature index (ANSI/UL 746B).*

Continuous using of temperature ranges or relative temperature index (RTI) is the maximum service temperature at which all critical material properties remain within acceptable limits for an extended period of time. According to UL 746B standard, one material can be assigned three independent RTIs:

① electrical —by measuring the dielectric strength;

② mechanical impact —by measuring the tensile impact strength;

③ sockless mechanical —by measuring tensile strength.

These three properties were chosen as critical performance in tests because of their sensitivity to high temperatures in service.

The material long time thermal performance is tested in comparison of a second control sample for which the RTI has already been determined and which has shown good performance. The control sample is used because properties that deteriorate with temperature increasing are sensitive to the variables of the test program itself. The control sample is affected by the same factors specific combinations during testing, which provides a reliable basis for comparison with the tested material.

Ideally, long time thermal performance could be assessed by aging the tested material at ambient temperature over a long time. However, this is not rational for most reasons. Therefore, accelerated aging applies at significantly higher temperatures. During aging, tested and control samples are placed in ovens, which have a predetermined constant temperature. Tested and control samples are ejected at specified time points and then tested to basic properties. By measuring these three properties as a function of time and temperature, the "end of service life" can be calculated for each temperature.

This "end of service life" is determined as the time it took for the material to deteriorate by 50% from the baseline. The maximum temperature at which the tested material will have a satisfactory service life can be determined by applying the test data into the Arrhenius equation. This calculated temperature is the RTI for each material property.

The RTI allows the engineer to use it to predict how polymer parts will perform in real application when exposed to elevated temperatures.

7) *Coefficient of linear thermal expansion (ASTM D696, DIN 53752).*

Each material expands when heated. Injection molded polymer CM parts expand and resize in temperature rise proportion. Researches use the coefficient of linear thermal expansion (CLTE), for estimation such expansion. It measures changes in the length, width, and thickness of a molded part.

Amorphous polymers are generally characterized by consistent expansion rates across their practically service temperature range. Crystalline polymers generally have increased expansion rates at temperatures above their glass transition temperature.

The CM reinforcement that creates anisotropy significantly affects the CLTE of the polymer. Glass fibers are usually oriented in the flow front direction: when the polymer is heated, the fibers prevent expansion along their axis and reduce the CLTE. In flow and thickness perpendicular directions the CLTE will be higher.

Polymers can be product with certain CLTE with accordance of the metals thermal expansion or other structural materials used in composite structures.

3. Flammability tests

The most widely accepted standards for flammability performance are UL94 (Insurance Research Laboratories) for polymers. These categories define the material ability to extinguish a flame after ignition. There are several categories based on burning rate, extinction time, drop formation resistance and whether the drops are flammable or non-flammable. Each test material can be marked by several categories depending on color and/or thickness. The UL category should be determined by the polymer part thinnest section for a particular material selection for the certain application. The UL category must always be marked with the thickness: simply listing the UL category without thickness is not enough.

1) Standard UL94HB categories classification

If flammability is a safety requirement, then the category HB (Table 7.11) materials using is usually not allowed. In general, materials HB categories are not recommended for electrical applications, except for mechanical and/or decorative parts. Non-fire-resistant materials (or materials that are not referred to as fire-resistant) automatically do not meet the category HB requirements. UL94HB is a flammability category and must be verified through testing.

Table 7.11　Standard UL94HB categories classification

Category	Flame properties
V-0	the vertical sample burning stops within 10 s; no drop formation is allowed
V-1	the vertical sample burning stops within 30 s; no drop formation is allowed
V-2	the vertical sample burning stops within 30 s; flammable drop formation is allowed
5V	the vertical sample burning stops within 60 s after fives flame action with the duration of each action to the tested sample for 5 s
5VB	wide plate samples can burn through and form holes
5VA	wide plate samples must not burn through (i. e. no holes)—this is the toughest UL category

For the horizontal sample arrangement (Figure 7. 27), and its slow burning:

① burning rate less than 76 mm per minute and less than 3 mm thickness;

② burning rate less than 38 mm per minute with a thickness of more than 3 mm.

 • *UL94V0, V1, V2 category*

For testing vertical samples are used the same samples as for HB test (Figure 7. 28).

All parameters are recorded: burning time, smoldering time, the drops appearance moment and ignition (or non-ignition) of the cotton lining. The difference between V1 and V2 is burning drops, which are the main source of flame or fire propagation.

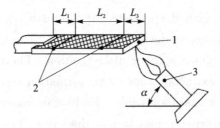

$L_1 = 25.4$ mm; $L_2 = 76.2$ mm; $L_3 = 25.4$ mm; $\alpha = 45°$

1—sample; 2—guide-mark; 3—flame source

Figure 7. 27   Horizontal sample flammability test layout

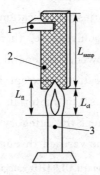

$L_{samp} = 127$ mm; $L_{fl} = 19$ mm; $L_{cl} = 9.5$ mm

1—support; 2 —sample; 3—flame source

Figure 7. 28   Vertical sample flammability test layout

 • *UL94-5V category*

UL94-5V is the most stringent of all UL categories. The tests are carried out in two stages.

Stage 1: The samples are fixed vertically and each sample is acted by a flame five times with a flame height of 127 mm each time for 5 seconds. To comply with the test conditions, no samples shall burn with the appearance of a flame or smolder for more than 60 seconds after the fifth flame actions. In addition, burning drops are not allowed which will ignite the cotton lining under the samples. The whole procedure is repeated with five samples.

Stage 2: A wide plate of the same samples thickness is tested in a horizontal position with the same flame. The whole procedure is repeated with three samples. These horizontal tests define two categories: 5VB and 5VA.

① category 5VB allows through burnout (with holes);

② category 5VA does not allow through burnout (without holes).

The UL94-5VA test is the most stringent of all UL tests. Materials of this category are used for fireproof casings (Figure 7. 29).

For these cases with expected web thicknesses of less than 1. 5 mm, fiberglass-filled materials should be used.

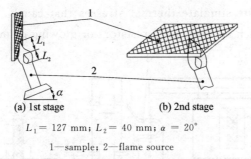

(a) 1st stage　　　　　　　　　(b) 2nd stage

$L_1 = 127$ mm; $L_2 = 40$ mm; $\alpha = 20°$

1—sample; 2—flame source

Figure 7. 29　UL94-5V category sample flammability test layout

2) *Flammability by Standard CSA C 22. 2 NO. 0. 6.*

These Canadian standard association (CSA) of flammability tests are similar to UL94-5V. But, these tests' conditions are stricter: each action of the flame lasts 15 seconds. In addition, during the first four actions of the flame, the sample should be extinguished within 30 seconds and after the fifth action within 60 seconds.

The results of these CSA tests shall be corresponded to the UL94-5V test results.

3) *Flammability index at limited oxygen content ISO 4589 (ASTM D 2863).*

The purpose of limited oxygen index (LOI) flammability determining is to measure the relative materials flammability when it burned in a controlled environment (Figure 7. 30). The LOI is the minimum oxygen content in the atmosphere that can provide a flame on a thermoplastic material. The test atmosphere is an externally controlled mixture of nitrogen and oxygen.

The fixed sample is ignited with an auxiliary flame, which is then extinguished. In successive test cycles, the oxygen concentration is reduced until the sample can no longer provide burning.

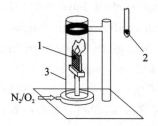

1—sample; 2—auxiliary flame; 3—controlled environment chamber

Figure 7.30   Limited oxygen index test layout

The LOI is determined as the minimum oxygen concentration at which a material can burn for three minutes, or can provide a burning propagation of a sample over a distance of 50 mm. The higher the LOI is, the lower the flammability is.

4) *Glow-wire test (IES 695-2-1).*

Glow-wire tests simulate thermal stresses that can be caused by a source of heat or ignition such as overloaded resistors or glowing elements (Figure 7.31).

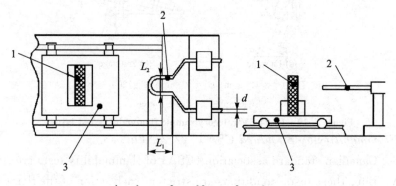

$d = 1$ mm; $L_1 = 12$ mm; $L_2 = 9$ mm

1—sample; 2—glow-wire; 3—movable support

Figure 7.31   Glow-wire test layout

A sample of insulating material is pressed for 30 seconds with a force of 1 N against the electrically heated glow-wire tip. The glow-wire tip motion into the sample is limited. After removing the wire from the sample, the flame extinguishing time and the presence of any burning drops are recorded.

The sample is considered to pass the glow-wire test when one of the following conditions occurs:

① In the absence of sample flame or smoldering;

② If the surrounding parts and the bottom layer of sample flame or smoldering goes out within 30 seconds after removing the glow-wire, the surrounding parts and the bottom layer would be not completely burnt out. In the case of using thin paper as the bottom layer, this paper should not catch flame, or there should be no scorching of the pine board, if used as the bottom layer.

Real live parts are tested in a similar manner. The glow-wire tip temperature level depends on how the tested part will be used in service:

① under monitoring or without monitoring;

② with or without continuous load;

③ located near or far from the central power supply point;

④ in contact with a live part or used as a casing or cover;

⑤ under less or more stringent conditions.

Depending on the required stringent level of the environment surrounding the part in real service, the following temperatures are preferred: 550, 650, 750, 850 or 960 ℃. An appropriate test temperature should be selected by assessing the failure risk due to impermissible heating, ignition and flame propagation.

5) *Needle-flame test (IES 695-2-2).*

The needle flame test simulates the effect of small flames that can occur due to faults inside electrical equipment (Figure 7. 32).

The test material layer is arranged under the sample to assess the possible flame spread (burning or smoldering particles) (Figure 7. 33). The test flame is applied to the sample for a specified period of time: usually 5, 10, 20, 30, 60 or 120 seconds. For specific requirements, other stringent levels may be adopted.

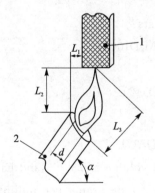

$L_1 = 5$ mm; $L_2 = 8$ mm;
$L_3 = 12$ mm; $d = 0.5$ mm; $\alpha = 45°$

1 —sample; 2 —flame source

Figure 7.32　Needle-flame test layout

Figure 7.33　Flammability test bench

In the absence of special requirements in the relevant specification, the sample is considered to have passed the needle-flame test if one of the following four situations occurs:

① If the sample is non-flammable.

② If the burning duration is not more than 30 s.

③ If the burning propagation specified in the test requirements has not been exceeded.

④ If a flame, burning or smoldering particles falling from the sample cause the fire to spread to the surrounding parts or to the layer placed under the sample, and if there is no flame or smoldering on the sample at the finish of exposure to the test flame.

4. Electrical tests

There are the following polymers electrical tests:

1) Dielectric breakdown strength IEC 243-1;

2) Area specific resistance IEC 93 (ASTM D257);

3) Volume resistivity IEC 93 (ASTM D257);

4) Specific inductive capacity IEC 250;

5) Dissipation coefficient IEC 250;

6) Tracking resistance ASTM D495;

7) Comparative tracking index (CTI) IEC 112.

5. Optical tests

There are the following polymers optical tests:

1) Milkiness and optical transmission ASTM D1003;

2) Politure DIN 67530 (ASTM D523);

3) Index of refraction DIN 53491 (ASTM D542).

6. Physical tests

1) *Density ISO 1183 (DIN 53479, ASTM D792).*

Density is the mass divided by the material volume at 23 ℃, and has usually the unit: grams per cubic centimeter ($g \cdot cm^{-3}$) or grams per milliliter ($g \cdot mL^{-1}$). "Specific density" is the ratio of the mass of a given material volume to the mass of the same water volume at a specified temperature. Density can be measured by several ways, as described in the ISO 1183:

① Method A. Immersion way of ready-made plastics.

② Method B. Pycnometer test for plastics in the form of powders, granules, tablets or molded parts reduced to small particles.

③ Method C. Titration way for plastics with shapes similar to those which are required for way A.

④ Method D. Density-gradient way for plastics similar to those required for Method A.

Density gradient columns are liquid columns whose density increases uniformly from top to bottom. They are especially suitable for measuring the small samples density and for comparing densities.

2) *Water sorption ISO 62 (ASTM D570)*

Plastics absorb water. Moisture content can change dimensions or properties such as insulation resistance, dielectric loss, mechanical strength and external view.

A certain size plastic sample water sorption determination is carried out by sample immersing in water for a specified time and at a specified temperature. The measurement results are expressed either in absorbed water milligrams or as a percentage mass increasing. It is possible to compare the different plastics water absorption when the tested samples are similar size and are in the same physical state.

The tested samples are preliminarily dried at 50 ℃ for 24 hours, cooled to room temperature and weighed before being immersed in water at a specified temperature for a specified time.

Water absorption can be measured:

① at 23 ℃. The samples are placed in a tank with distilled water at a temperature of 23 ℃. After 24 hours, the samples are dried and weighed.

② at 100 ℃. The samples are placed in boiling water for 30 minutes, cooled for 15 minutes in water at 23 ℃ and weighed again.

③ until saturation. The samples are immersed in water at a temperature of 23 ℃ until they are completely saturated with water.

Water absorption can be expressed as:

① mass of absorbed water;

② mass of absorbed water per unit surface area;

③ percentage of absorbed water in relation to the tested sample mass.

7. Viscosity tests

1) *Molding shrinkage ISO 2577 (ASTM D955).*

The molding shrinkage is the difference between the dimensions of a mold and a molded part obtained in that mold. It is recorded in % or millimeters per millimeter.

The molding shrinkage are recorded both parallel to the material flow (in the flow direction) and perpendicular to the flow (in the cross-flow direction). For fiberglass materials, these values can vary significantly. Molding shrinkage can also vary from other parameters, such as part and mold structures, mold temperature, specific injection pressure, and molding cycle time.

Molding shrinkage (when measured on simple parts such as a tensile test sample or disc) are only typical data for material selection. They cannot be applied to structural parts or units.

2) *Melt flow rate or melt index ISO 1133 (DIN 53735, ASTM D 1238).*

Melt flow rate (MFR) or melt flow index (MFI) tests measure the molten polymer flow through an extrusion consistometer under specified temperature and load conditions (Figure 7.34).

The extrusion consistometer consists of a vertical cylinder with a small 2 mm diameter extruder head at the bottom and a removable force plunger at the top. The material portion is placed in a cylinder and preheated for several minutes. A plunger is mounted on the top surface of the molten polymer and its weight pushes the polymer through the extruder head onto the plate. The test period varies from 15 seconds to 6

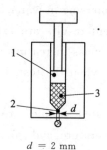

$d = 2$ mm

1—force plunger; 2—extruder head; 3—polymer

Figure 7.34　Melt flow rate test layout

minutes depending on the polymer viscosity. The applied temperatures are 220, 250 and 300 ℃. The weights of the applied loads are 1.2, 5 and 10 kg.

Laboratory equipment is shown in Figure 7.35.

Figure 7.35　Laboratory equipment for measuring melt index

The polymer amount collected after a given test period is weighed and converted to the number of grams that could be extruded after 10 minutes. The MFR is expressed in grams per reference time.

Example: MFR (220/10) $=xx$ grams / 10 minutes. It means the MFR at a test temperature of 220 ℃ and a rated load mass of 10 kg.

The polymers MFR depends on the shear rate. The shear rates in these tests are significantly lower than those used under normal manufacturing conditions. Therefore, the data obtained by this way may not always correspond to their properties in actual application.

　3) *Melt volume index ISO 1133 (DIN 53735, ASTM D 1238).*

The DIN 53735 standard describes three ways for the flow measuring:

　① Verfahren A;

　② Verfahren B, which in turn includes two ways:

　③ Mebprinzip 1;

　④ Mebprinzip 2.

The Verfahren A way consists of mass measuring when polymer is extruded through a given head.

The Verfahren B way measures plunger displacement and material density under similar conditions.

The Verfahren B/Mebprinzip 1 way measures the plunger displacement.

The Verfahren B/Mebprinzip 2 way measures the plunger moving time.

The flow index according to Verfahren A standard DIN 53735 is equal to the MFR according to the ISO 1133 standard.

The standard DIN 53735 describes the melt volume index (MVI). The ISO 1133 standard does not have the MVI. The MVI is defined as the polymer volume that is extruded through the extruder head for a given time. The MFI is defined as the polymer weight extruded through the extruder head for a given time. The MVI is expressed in cm^3 per 10 min and the MFI in gram per 10 min.

The temperatures are applied 220, 250, 260, 265, 280, 300, 320 and 360 ℃. The applied load weights are 1.2, 2.16, 3.8, 5, 10 and 21 kg.

Example: MVI (250/5) means the melt volume index in cm^3 per 10 minutes for a test temperature of 250 ℃ and a nominal weight of 5 kg.

　4) *Melt viscosity DIN 54811.*

The melt properties are determined in a capillary viscometer. There are measurements:

　① the pressure at a given volumetric flow rate and a given temperature;

　② the volumetric flow rate at a given pressure.

Melt viscosity (MV) has Pa · s unit.

　5) *Practical application of MV, MFR/MFI, MVI in a polymer production.*

The MV test with measurement in a capillary viscometer is very similar to the normal extrusion process. As such, the MV test is a good basis for comparing the flow of injection-molded materials: it shows the pass melt viscosity through the nozzle.

The MFR/MFI and MVI tests, where the shear rate is too low, are not suitable for application in the injection molding process. They are very good data for the inspection by the manufacturer: easily, quickly and inexpensively, but not suitable for material selection in terms of its expected molding flow.

7.2.3　Composite material environmental certification

The environmental certification goal is to stimulate manufacturers to introduce such technological processes and develop such products that pollute the environment to a minimum and give the consumer a guarantee of product safety for his life, health, property and habitat, and accelerate the products promotion on the market.

Special attention should be paid to the polymer and polymer-containing materials production.

The most common sources of negative impact on the human health and the environment are polymeric materials and products (for example, building materials). In addition to the polymer, which is a binder, the composition of plastics includes:

1) reinforcements in the form of organic or mineral powders, fibers, threads, fabrics, sheets;

2) plasticizers;

3) stabilizers;

4) curatives;

5) dry colors.

Plastics have a number of important physical and mechanical properties, which determine the main areas of their application in engineering. Today on the polymers basis, a lot of structural and decorative materials are made.

The most common polymers used in the building materials production are:

1) by the polymers class obtained by chain polymerization: polyethylene, polypropylene, polyvinyl chloride, polyisobutylene, polystyrene, polyvinyl acetate, polyacrylates and coumaroneindene polymers;

2) by the polymers class obtained by polycondensation and stepwise polymerization: phenolic, phenol-formaldehyde, resorcinol-formaldehyde, polymers based on amido and amino-formaldehyde polycondensation, glyphthalic polymers, polyurethanes, polyester maleinate and polyether acrylate and epokaldehyde, silicon polymer and

epoxide polymers.

For building plastics and other synthetic-based materials for indoor application, toxicological and hygienic properties are the most important.

The materials' groups that are dangerous to human health include plastics that emit toxic substances in quantities exceeding the standard values. For example, as a result of incomplete chemical processes of polymerization/polycondensation, plastics can release: acetone, benzol, phenol, furfural, chlorine, vinyl acetate, etc. Even small concentrations of these substances can cause disturbances in the human health. The most common causes of allergic dermatitis and eczema are formaldehyde, epoxy, polyester resins.

Some polymer materials for floors (linoleums, polymer tiles, carpets), laminated plastics, fiberboards, synthetic wallpaper, various polymer additives in concrete and matrixes, synthetic adhesives, synthetic-based insulation, etc. can serve as a source of formaldehyde. Moreover, materials with a protective layer will emit formaldehyde longer than materials without a protective layer.

The substances emission rate also depends on temperature, humidity and the quality of air exchange.

Linoleums can release phenol, benzol; finishing materials based on polystyrene can release ayrol. Synthetic wallpaper can be an emission source of free vinyl acetate, ayrol, vinyl chloride, acrylic, etc. Benzol, hexene, toluene, methylene, ethyl benzol and other toxic compounds can emit into the environment when PVC products used.

Polyurethane foam dust irritates the skin and eyes mucous membranes, and when it enters the lungs, it reacts with the protein, gradually changes its structure and promotes the emphysema development.

The composition of the "dangerous mixture" that is formed by gaseous products emission depends on the production approaches of synthetic materials and their operation modes.

Substances, released during the plastics burring, can be extremely toxic. So, when PVC burns, dioxin is formed. It is the strongest poison, an insignificant dose of which is fatal to humans. It is a thousand times more toxically than the well-known strychnine poison. Dioxin has a dangerous effect on the human's immune and endocrine systems. It is also causing the oncomas formation and negatively affects reproductive functions. This poison action presents even at extremely low concentrations. In Russia, the dioxins admissible daily dose is 10 picogram per kg of body weight (a picogram is one trillionth of a gram).

It should be noted that static electricity accumulates on plastic surfaces, which not only negatively affects human heart and nervous activity, but also attracts synthetic dusty particles. And dust is a favorable environment for microbes.

In principle, the environmental certification practice of polymer products shows that with a correctly selected production technology, the toxic compounds emissions can be negligible.

7.3 Certification approach for an aircraft with composite material units

The base for the aircraft certification is airworthiness regulations. Providing satisfactory airworthiness requires that contradictory structural properties be balanced to increase the aircraft safety level and is a key factor in design. At the early aviation industry stages, aircraft designing paid attention to developing excellent aerodynamic and structural strength performance. A "safe" aircraft was by-product. Modern design approaches content airworthiness requirements among flight performance, operational effectiveness, reliability, maintainability, and economy. All of them are important in determining the aircraft success.

The important step is delivery of the final product and the acceptable safety level demonstration. In the metallic structures case, this task has been solved by certification procedures. They are based on service experience and science approaches. The safety factor that applied to full-scale test results is an example of such science and experience combination.

A clear analytical approach based on the reducing test-demonstrated-lives and in-service-lives with an acceptable failure low possibility would show a factor about three. But, through aircraft application data long term research and engineering approaches, this safety factor value was proofed of two ($f = 2$). Because CM structures are relatively new in the aviation industry, the certification process development should base less on experience and more on the CM behavior.

Already existing requirements for metals were adopted for initial CM structure certification attempts. They were acceptance without paying attention to the differences between the two materials, even though these differences can significantly affect safety. For example, CM usually show linear elastic behavior till failure and are extremely sensitive to stress concentrations under static loading. But, metals, with some exceptions, show plastic behavior above a yield stress and are not notch sensitive under static loading. Another example of significant differences between composite and metallic structures is in their failure durability under compression. CM structures are much more failure sensitive and for this reason there are strict requirements on strength for composite parts. Usually, certification procedures relate with the failure durability in composites.

Additively, another critical difference is damage growth due to fatigue. It is often a critical condition for metals, whereas CM usually show high resistance to such loading. The stress levels related with structural critical load cases in CM, such as compression together with impact damage, have typically been low enough to ensure that the damage does not grow due to fatigue. Thus, CM structures have usually been depended on static conditions rather than by fatigue. As designers want to fully apply the CM specific strength and stiffness, the stress levels will increase, and fatigue must necessarily be given greater consideration in the certification of new aircraft.

Perhaps the most important difference between CM and metals is in their varying properties under different operational conditions. The CM structures regression in certain environmental conditions was led to some standard certification approaches. Essentially, it is need to determine the critical material properties after extreme thermal and moisture exposure in which the structure will operate. In addition, it must be shown that there would be no regression after chemicals exposure that can be present in real operation (hydraulic oil, fuel, de-icing fluids, etc.).

CM based on thermosetting resins usually show significant sensitivity to absorbed moisture and temperature. The moisture level in thermosetting resins is proportional to atmospheric humidity. Most structural CM (usually epoxy based) can absorb up to one percent by moisture weight in a typical aircraft operational environment. The moisture softens the matrix resin and reduces those properties that depend on the resin, such as shear and compression stiffness and strength. This affect is mostly evident at high temperatures, when the resin is additionally softened. The simultaneous applying of environmental effects with structural testing especially of full-scale structures is very expensive. An alternative approach to account for the CM properties decreasing at high temperature and humidity is to apply the increasing loads in certification tests conducted at ambient temperature and humidity.

Unluckily, there are additional problems of combined metallic and composite structures, because the stress factor for the CM structure is higher than for metallic components.

The Federal Aviation Authority (FAA) has developed a document, FAA AC 20-107A 7 that explains acceptable ways of demonstrating compliance to the airworthiness certification requirements for CM structures. The document has the additional approaches that must be applied specifically to the CM structures certification:

1) the operational environment effect on material properties and its manufacturing process;

2) static strength in terms of operational conditions, repeated loading, impact failure and material variability;

3) fatigue and fail-safe estimation.

There are also some additional requirements that included flutter, flammability, lightning protection, quality inspection, maintenance, and repair.

In the document AC 20-107A, the critical task of correct approval environmental conditions is related to allowing either full-scale testing or testing through a "building-block" method. This method is the most often used, because of its lower cost. But, it involves wide engineering development testing to:

1) obtain environmental and scatter reduction (or load adjusting) factors for critical failure modes;

2) verify critical structural specifics.

Full-scale testing will also be required under different environmental conditions.

The major problems that affected CM structures are the static and fatigue strengths, environmental effect, fail-safe and flammability.

7.3.1 Airframe structures certification

The next main requirements have been developed on the metallic airframes experiences base and continue to be the basis for the CM airframe structures certification. The structure (by test and/or analysis) shows the following abilities:

1. Static strength

1) Design limit load, no failure or unacceptable deformations. Design limit load is typically the maximum load expected to be applied on the structure in its service life.

2) Design ultimate load, no failure, limited residual deformation is acceptable. It is typical that design ultimate load is higher than design limit load at 1.5 time.

2. Fatigue strength

1) Safe life approach. No substantial cracking that could lead to failure should happen in the airframe service. This approach was applied for most of the old metallic structure aircraft.

2) Fail-safe approaches:

① Alternate load-path. The structure is fail-safe in that meaning: cracking may happen, but will not reduce strength lower than an acceptable level before being detected. This property is generally met by multi-load-path structure. If one load path fail, other load paths can continue to support the required permanent strength level before the damage is detected. This approach is widely used in the civil aircraft structure.

② Slow crack growth approach. The structure is fail-safe in that cracking may happen. But cracks will grow slow and will not reason failure for the full structure service life or will not reason failure before detection in time planned inspection (safety by inspection). This approach can be used to single-load-path structure, where failure would be critical. Fail-safe single-load-path structure is based on the supposed defect's presence at critical zones.

3. Fail-safe general requirement

The strength will not decrease below an acceptable level (usually at 20% higher than design limit load) due to typical structure failure (caused by fatigue cracking, corrosion, random mechanical contact, etc.) until being detected. Critical failure must be of a size that can be detected with a high probability rate.

4. Durability and economic requirements

For the airframe service life, failures that require costly repairs will not happen (for example, due to fatigue or corrosion). This is not ordered airworthiness requirement that approved by the airworthiness agency. However, it is an important approval in the commercial aircraft selection.

5. Civil aircraft test program

CM structures are required to show the same safety level as a regular metallic structure. But, the compliance means has to take into account the differences in material behavior.

Airframe design for static strength consists of:

1) a detailed structural analysis, typically a structural finite-element simulation applied;

2) a samples test program of increasing complexity, from simple tested samples to assembled structural elements or full-scale structures, as shown in Figures 7.3 and 7.36.

Sample and structural members tests are used to receive material and structural allowance for structure and must therefore examine all critical loading and possible failure modes. The other tests are mainly for structure proof.

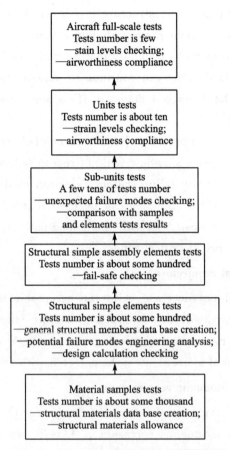

Figure 7.36   Civil aircraft test program

7.3.2   The design allowance development

Design allowance have to be set at the most critical environmental conditions. "Cold-dry" and "hot-wet" extremes are the most critical. Because tested samples and structural elements are usually small, there are no difficulties in moisture them. To ensure objectivity, airworthiness regulations require full moisture-saturation at the highest expected operating temperature. Most attention must be given to matrix-dominated failure modes, because they are most sensitive to degradation.

CM samples tests are carried out on their lay-outs base to determine the allowable critical "temperature-moisture" combinations. These tests determine design allowance and reduction factors. A comparison of the room "room temperature-dry" and the "hot-wet" conditions gives an environmental reduction factor. Similarly, any mean value comparison with its related allowable value provides a reduction factor.

Some structural element tests are made to verify allowance at the badly environmental conditions. These tests typically include "open-hole", "filled-hole" tension and compression, including load bypass. Tests should also include details of out-of-plane loading zones, such as stiffener run-outs.

The design allowance development on the base of samples and structural elements tests is a very important, complex and expansive design and certification process part. This is especially typically for CM because of the mechanical properties high scatter.

1. Static strength allowance

Airframe static strength design is based on sample and structural element data that allows statistical changes or scatter in strength. Two allowable statistical levels are possible:

1) A-allowable—value achieved by 99% of the amount at the 99% trust level;

2) B-allowable—value achieved by 90% of the amount at the 95% trust level.

For this allowance determination, the statistical model that best corresponds the property distribution is early determined. For structural metals, the property distribution is typically normal or (for fatigue) log-normal. For CM, the first model estimated is typically the two-parameter Weibull distribution. It has a more physically abstinent model for brittle materials.

Testing economic aspect is important. It based on the samples minimum number determination that need to be tested to obtain acceptable allowance. It depends on the statistical features. Tests number can be reduced significantly if the distribution features are already known. For low scatter structural materials, such as airframe alloys, the A-allowable strength is typically applied because this patently provides the highest safety margin. But, in cases of large scatter structural materials, this may lead too high and useable strength penalty. The B-allowable strength is appropriate for fail-safe or multiple-load-path structure. For CM, the B-allowable is typically applied because of the relatively high strength scatter.

If service conditions can lead to a further static strength reduction, the allowable static strength may be decreased by reduction factors applying. A same procedure can be applied to determine structural element allowance. Some reduction factors can be used to sample data to determine the final design allowance.

This approach avoids comprehensive tests that would be needed to allowance determining for all possible conditions and structures. It is usually supposed that the scatter is unchanged from that obtained when determining the allowance.

2. Fatigue allowance

The fatigue or durability allowance determination, CM samples and structural elements are tested under constant amplitude and also under spectrum loading typical of expected service conditions. The most environmentally degraded condition simulation is conducted at high temperatures with the sample accordingly moisturized.

The fatigue tests under constant amplitude show basic data for evaluation of spectrum loading behavior and set load ignore levels. The load ignore levels are those can be neglected in the test. Additionally, constant amplitude tests show environmental effects.

The main fatigue tests feature is that results scatter is significantly higher than for other mechanical properties. So, a lot of these long time tests are needed to obtain the allowable fatigue values. Fiber CM have a higher stress versus cycles number to failure curve compared with structural metals because they are highly resistant to fatigue.

Significant damage growth typically happens only at strain levels above 60% of the static strength. But, once growth starts, its development is very rapid, often catastrophic. Thus, in contradistinction from metals, the slow damage growth mode for CM is not considered possible, therefore real damage growth data are infrequently obtained.

Same situation presents for bonded joints for which the damage growth rate can also be fast, and tests are typically made to obtain the damage growth limit.

It is usually found that when the various reduction factors are used to the static strength received from the sample data, the final design allowance are enough to provide a sufficient in-plane strength degradation margin under cyclic loading. But, this may not be enough to evaluate for strength degradation in some joint types.

3. Damage influence allowance

For fail-safe structure, the strength allowance reduction due to impact damage, over the likely energy levels range, must be evaluated. Cyclic loading at typical strain levels can be reason for further loss in allowable static compression strength and limited damage growth and therefore will also require estimation.

It is important in the design allowance determination to evaluate the impact damage influence on structural elements such as skin panels with reinforcement because the damage location is very important. If impact damage is located between reinforcement, failure can happen in two steps if the damage exceeds a limit. The first is rapid damage growth till the reinforcement. The next with delay is reinforcement failure.

7.3.3 Static strength demonstration

The same test program for a metal structure is usually applied in a CM structure. But, it is needed to take into account the significant differences. A large number of sample and elements tests are completed to obtain general design data and to evaluate the operational environment and damage effect on the materials and elements used in the structure. They are also applied to check safe strain levels in sub-units, units and the full-scale tests.

Smaller sub-unit number tests (see Figure 7. 36), are used to adjust the forecast made for the full structure from the element and sample test results and to verify critical structural features. Certification typically bases on some regular-temperature ambient, full-scale static and fatigue tests. Environmental effects are evaluated in either the tests or the tests result analysis. Such "building-block" method to CM certification is typically applied in commercial purpose.

1. Structural element and sub-unit tests

Structural element and sub-unit tests are made to develop specific data related to the given design. The structural elements and sub-units selected for test will initially be based on the finite element model forecasts. After that, they are tested to failure under hardest environmental conditions and the failure strain determined. These tests are given the ultimate strains average values in the environmentally badly condition. Because of the size limits of sub-units, these are often tested at ambient conditions and reduction factors used from sample tests. The sub-units failure zone and nature must be the same with the structural element tests and shown the same failure mode. If this is not the case, then such tests must be repeated with adjustments to the loading or limits. Then, expecting that the structural element and sub-units tests scatter are the same as those in the element tests, the reduction factor application to the mean values gives allowable values for the full-scale structure in the environmentally badly condition.

An important economic question is the time required to develop an adequate moisture distribution. Depending on the thickness of the CM structure, this may take from some weeks to some months.

2. Full-scale tests

The full-scale test is important to research the secondary loading effect caused by out-of-plane loading. Such loads occur from offsets, stiffness varies, discontinuities, and local buckling, which can not be fully predicted or excluded in design or determined by the structural element sample. Additionally, it is also important to verify the finite element model to ensure that structural internal load arise as forecasted.

The full-scale unit or structure, which may be a wing, fuselage, or full aircraft, is typically tested in the "room temperature-dry" condition. The basic difference between this test on a composite and a metal aircraft is that, for the CM, the structure is much more widely strain-evaluated. This test most serves to verify the finite element model. If the strain evaluated results show zones of high strain in zones where no sub-units were tested, then it is needed that such tests completed. Then, based on the ultimate load test, the measured strains at 150% design limit load are compared with the reduced design allowance as set by the sample and structural element tests. If the measured strain is higher than the allowable value, failure is considered to have happened, and some structural adjustment is necessary. There is indeterminacy at various test stages, but the general approach is reasonable. It can be seen that, for CM aircraft, actually all design testing (on small and large samples) gets an integral part of the airworthiness certification.

But, it is not the single approach to showing static strength. One alternative seldom proposed is to carry out the full-scale structure test under ambient conditions, but with the used loads increased to allow for environmental effects. The load increase is determined from sample tests. Another alternative is to perform a full-scale environmental test.

3. Proof tests

In the past, because of certification or non-destructive inspection problems, airworthiness agencies have required that every produced CM element be given a proof test, typically to a load slightly in excess of the design limit load. In such cases, the elements are given a thorough non-destructive inspection both before and after test to check for failure that may be caused by the proof test. Because assurance in the material properties and analysis approaches has improved, this way is no longer usually insisted upon.

7.3.4 Fatigue strength demonstration

The composite aircraft structure fatigue performance tests are far from clear demonstration. At least one requirement is to check for through-thickness failure modes, caused, for example, by unexpected through-thickness stresses.

The full-scale fatigue tests that have been performed on aircraft containing CM have typically been the same as those would have been performed for a traditional metal aircraft. In other words, a test is to N lifetimes (where N may be 2, 4, etc.) in a standard environment. The fatigue test on a full-scale CM aircraft with the specific "moisture-temperature" environment is complex and expensive. One way test for wings with integral fuel tanks is to fill them with heated water. However, this way is not applicable for most other structures. Also, adequate data are not present on the fatigue state of large-size CM structures to apply any real scatter factor. The normal environment full-scale tests will of course continue to be applied to the metal structure verification, and they may sometimes also apply to observe unsuspected problems with the CM structures. But, the major CM structures fatigue performance validation will be based on structural element, sub-unit and unit testing in a required humid environment, with the certain temperature cycling (especially the thermal extremes) precise conception.

The obvious composite fatigue life wide scatter would show the necessary for CM structures testing for overmuch long time (about 30 or more service lifetimes) to adequately account for the changeability reduction factor. By certain loads elevation, it is possible to do the test to be spent for only one service lifetime. If this is too hard on other airframe structural parts, the using of partly reduced loads can provide the tests to be spent to only two or three service lifetimes. An important problem in combined composite and metal airframe structure is that the metal parts may need to be reinforced (over that the actual airframe requires) to provide sufficient CM parts testing under elevated loads.

The CM fatigue is surely a reason for major troubles. It is more a problem of there being difficulties in determination an adequate test demonstration.

7.3.5 Damage tolerance demonstration

Damage tolerance is usually estimated at all test levels: from the sample to the full-scale structure. Damage can be advisedly created, for example, applying a portable impact tester in dangerous zones, such as joints. Failure can also be cuts, disbonds or delamination occurred in time of manufacturing.

Damage tolerance demonstration on full-scale structures includes residual strength inspection typically performed after some service lifetimes of fatigue cycling have been shown on an undamaged unit. Damage is then created on the tested structure in the critical zones and limit-load-testing carried out. Usually, at this step, the fatigue testing is continued for one or more next service lifetimes to get any failure growth. If that damage has not happened, the next stage is to maintain the unit and continue fatigue cycling for a next one or two service lifetimes. So, the structure is tested to damage. An acceptable, if enough conservative, result would be to obtain ultimate load in the test.

7.3.6　The impact damage threat assessment

For damage tolerance estimation and for the designing process, it is important to evaluate the impact damage threat for the airframe structure. Of course, surfaces that are high and vertical surfaces on an aircraft have less dangerous to damage than horizontal surfaces. The upper horizontal surface will be more affected to damage from hailstones or dropped tools, but the lower surface will be more affected to damage from runway foreign objects. The measured damage depth is transfered, by calibration to the equivalent strike energy. Then, the threat is quantified by the Weibull distribution. It predicts the strike occurrence probability over the energy levels range. This data could be applied to plan the damage's level and location on a test structure.

Questions

(1) Describe the structural material selective criteria.

(2) What are the features of composite materials as structural material?

(3) How are you understand "Building block" approach?

(4) Describe the material failure rate briefly.

(5) Describe the manufactural defects and quality inspection briefly.

(6) Describe the requirements for regulatory and engineering support for certification briefly.

(7) Explain the criteria for selecting the design solutions for composite material structure.

(8) Explain the general scheme for polymer composite materials application effectiveness increasing.

(9) What is composite materials structure operational survivability depending on?

(10) What is the certification process?

(11) What are the thermal behavior indexes of composite materials?

(12) What are the chemical fastness indexes of composite materials?

(13) What are the electrical properties indexes of composite materials?

(14) Describe the polymer flammability properties of composite materials.

(15) Describe the tribological properties of composite materials.

(16) Briefly explain mechanical tests ways for composite materials.

(17) Briefly explain flammability tests ways for composite materials.

(18) Briefly explain physical tests ways for composite materials.

(19) Briefly explain viscosity tests ways for composite materials.

(20) Describe briefly the composite material environmental certification.

(21) Explain Civil aircraft test program.

(22) Describe briefly the design allowance development.

(23) Describe briefly the fatigue strength demonstration.

REFERENCES

[1] Atroshchenko E S, Rosen A E , Golovanova N V. Development of scientific foundations for the formation of structure and properties of composite materials[J]. Materials Science, 1998(6).

[2] Basakov M I. Fundamentals of standardization, metrology [M]. Rostov: Phoenix, 2005.

[3] Bulanov I M, Sparrow V V. The technology of rocket and aerospace structures made of composite materials[M]. Moscow: MSTU Bauman, 1998.

[4] Cardrick A W, Curtis P T. Certification of Composite Structures for Military Aircraft[J]. D/RAE(F) M&S New Materials and Processes, 1992,1(5): 121-135.

[5] Mikino A, Odawara O, Miyamoto E, et al. Chemistry of synthesis by combustion[M]. Moscow: Mir, 1998.

[6] Collins R. Fluid flow through porous materials[M]. Moscow: Mir, 1964.

[7] Karpinosa D M. Composite materials[M]. Kiev: Naukova Dumka, 1985.

[8] VasilievaV V, Tarnopolsky Y M. Composite materials[M]. Moscow: Mechanical Engineering, 1990.

[9] Vasiliev V V, Protasov V D, Bolotin V V, et al. Composite materials[M]. Moscow: Mechanical Engineering, 1990.

[10] Brautman L, Kroc R. Composite materials: in 8 vol. [S]. Moscow: Mechanical Engineering, 1978.

[11] Brautman L. Destruction and Fatigue[S]. Moscow: Mir, 1978: 486.

[12] Dimitrienko Y I. Mechanics of composite materials at high temperatures[M]. Moscow: Mechanical Engineering, 1997.

[13] Dmitrov L N, Kuznetsov E V, et al. Bimetals[M]. Perm, 1991.

[14] Dovgyalo V A, Yurkevich O R. Composite materials and coatings based on dispersed polymers[M]. Minsk: Navuka i tekhnika, 1992.

[15] Drozhzhin A V. Influence of the surface structure of polymeric composite materials on the strength of the repair adhesive bond for improving the repair of aviation equipment[J]. Interuniversity. Sat. scientific papers, 1989.

[16] Dutton S E, Lofland R A. Certification of Structural Composite Components Used on the MD900 Helicopter[J]. Proceedings of ICCM-11, 1997.

[17] Ekhan H P, Cordero R, Whitehead R S. Advanced Certification Methodology for Composite Structures [S]. DOT/FAA/AR-96/111, Federal Aviation

Administration, 1997.

[18] Elby D, Hoffman P, Hoffman M, et al. Managing Safety: The Mathematical Basis and Origins of the Navy's Factor of Safety [J]. Naval Aviation Structural Integrity and Aging Aircraft Conference, 1998.

[19] Fawcett A, Trostle J, Ward S. 777 Empennage Certification Approach[J]. Proceedings of ICCM-11, 1997.

[20] Federal Aviation Administration. Composite Aircraft Structure[S]. Advisory Circular AC 20-107A, 1984.

[21] Federal Aviation Administration. Federal Aviation Regulations. Part 25. Airworthiness Standards: Transport Category Airplanes[S]. U. S. Department of Transportation, 1994.

[22] Ageeva N V, et al. Fibrous and dispersion-strengthened composite materials [M]. Moscow: Nauka, 1976.

[23] Bokstein S Z. Fibrous composite materials[M]. Moscow: Mir, 1967.

[24] Shorshorov M K, Kolpashnikov A I, Kostikov V I, et al. Fibrous composite materials with a metal matrix[M]. Moscow: Mechanical Engineering, 1981.

[25] Yanagida H. Fine technical ceramics[M]. Moscow: Metallurgy, 1986.

[26] Fistul V I. New materials. Condition, problems, prospects[M]. Moscow: MISIS, 1995.

[27] Goldstein M I, Litvinov V S, Bronfin B M. Metal physics of high-strength alloys[M]. Moscow: Metallurgy, 1986.

[28] Gulyaev A P. Metallurgy[M]. Moscow: Metallurgy, 1977.

[29] Holister G S, Thomas K. Materials hardened by fibers [M]. Moscow: Metallurgy, 1969.

[30] Metcalfe A. Interface surfaces in metal composites[M]. Moscow: Mir, 1978.

[31] Isaev L K. Metrology and standardization in certification[M]. Moscow: IPK Publishing house of standards, 1996.

[32] Ivanova V S, Kopiev I M, Elkin F M. Fiber-reinforced aluminum and magnesium alloys[M]. Moscow: Nauka, 1971.

[33] Kapustin A I, Nuzhdin G A. Obtaining and properties of superhard composites[M]. Moscow: Mechanical Engineering, 1999.

[34] Kardashov D A. Synthetic adhesives[S]. Moscow: Chemistry, 1976.

[35] Karpinos D M, Tuchinsky L I, Vishnyakov L R. New composite materials [M]. Kiev: Vishcha shkola, 1977.

[36] Kiparisov S S, Libenson G A. Powder metallurgy[M]. Moscow: Metallurgy, 1980.

[37] Konovalov B A. Residual three-layer shell made of a composite material with

a macrodefect in the outer layer of strength and durability of elements of aircraft structures[J]. Interuniversity. Sat. scientific papers, 1990.

[38] Korten H T. Destruction of reinforced plastics[M]. Moscow: Chemistry, 1967.

[39] Koshevaya I P, Kanke A A. Metrology, standardization, certification[M]. Moscow: INFRA-M, 2008.

[40] Krupin A V, Soloviev V Y, Popov G S, et al. Explosion treatment of metals [M]. Moscow: Metallurgy, 1991.

[41] Krylova G D. Fundamentals of standardization, certification, metrology[M]. Moscow: UNITI-DANA, 2000.

[42] Krysin V N, Krysin M V. Technological processes of forming, winding and gluing structures[M]. Moscow: Mechanical engineering, 1989.

[43] Lameris J, The Use of Load Enhancement Factors in the Certification of Composite Aircraft Structures[S]. NLR TP 90068U, National Aerospace Laboratory, 1990.

[44] Lifits I M. Fundamentals of standardization, metrology and certification[M]. Moscow: Yurayt, 2001.

[45] Marchenkov V I. Jewelry making[M]. Moscow: Higher school, 1992.

[46] Matvienko V A, et al. Using the method of multicriteria optimization to determine the parameters of the formation of the adhesive layer[J]. Aviation industry, 1992(5).

[47] McCarty J E, Johnson R W, Wilson D R. 737 Graphite-Epoxy Horizontal Stabilizer Certification[J]. AIAA 82-0745, 1982.

[48] Sendetsky J. Mechanics of Composite Materials[S]. Moscow: Mir, 1978.

[49] Meekoms K J. The Origin and Evolution of the Design Requirements for British Military Aircraft[S]. RAE publication, 1983.

[50] Naidich Y V. Contact phenomena in metal melts [M]. Kiev: Naukova Dumka, 1972.

[51] Non-destructive testing of permanent joints made of polymer composite materials[S]. Moscow: NIAT, 1994.

[52] Melnikov P. Reference book on electroplating in mechanical engineering[S]. Moscow: Mechanical Engineering, 1991.

[53] Goldberg M M. Paints and varnishes for mechanical engineering[M]. Moscow: Mechanical Engineering, 1974.

[54] Park W J. On Estimating Sample Size for Testing Composite Materials[J]. Journal of Composite Materials,1979(13).

[55] Perry F S. Harrier II : A Comparison of U. S. and U. K. Approaches to Fatigue Clearance [J]. AGARD Meeting on an Assessment of Fatigue

Damage and Crack Growth Prediction Techniques, 1993.

[56] Prilutskiy E V, Gridneva I V, Milman Y V, et al. About mechanical properties of silicon carbide fibers[J]. Moscow: Physics and chemistry of materials treatment. 1979(5).

[57] Lubin J. Reference book on composite materials[S]. Moscow: Mashinostroenie, 1988.

[58] Bunakov VA, Golovkin G S, Mashinskaya G P, et. al. Reinforced plastics [M]. Moscow: Publishing house MAI, 1997.

[59] Repair of honeycomb glued structures made of polymer composite materials [S]. Moscow: NIAT, 1984.

[60] Ripley E L, Cardrick A W. The UK Approach to the Certification of Composite Components for Military Aeroplanes and Helicopters[S]. AGARD Report No. 660, 1977.

[61] Semenov B I, Romanova V S. Aluminum foamed composites are a promising class of new materials for mechanical engineering and instrument making [J]. Materials Science, 1998(9).

[62] Shayakhmetov U S. Composite materials based on silicon nitride and phosphate binders[M]. Moscow: SP Intermet Engineering, 1999.

[63] Shvedov S. Short course on standardization, metrology and certification[M]. Moscow: Okay-kniga, 2007.

[64] Sidorov A I. Restoration of machine parts by spraying and surfacing[M]. Moscow: Mashinostroenie, 1987.

[65] Sirotkin O S, et al. Influence of elastic parameters of composite materials on stress concentration in the zone of stitched elements[J]. Aviation Industry, 1989(5).

[66] Smyslova R A, Kotlyarov S V. Reference book on sealing materials based on rubbers[S]. Moscow: Chemistry, 1976.

[67] Solntsev Y P, Pryakhin E I, Voytkun F. Materials science[M]. Moscow: MISIS, 1999.

[68] Stepanov A A. Cutting of high-strength composite materials[M]. Leningrad: Mechanical engineering, Leningraddepartment, 1987.

[69] Sutton W, Chorns J. Prospects for metals reinforced with fiber oxides[J]. Fibrous composite materials, 1967.

[70] Tailor K I, Babich B N, Svetlov I L. Nickel-based composite materials[M]. Moscow: Metallurgy, 1979.

[71] Tailor K I, Zabolotsky A A, Salibekov S E, et al. Classification of composite materials[J]. Powder metallurgy, 1977(12).

[72] Tarnopolsky Y M, Zhigun I G, Polyakov V A. Spatially Reinforced Composite Materials[M]. Moscow: Mechanical Engineering, 1987.

[73] Potapov A I, Ignatov V M, Aleksandrov Y B, et al. Technological non-destructive testing of plastics[M]. Moscow: Chemistry, 1979.

[74] Ivanov VS, s: dorov RT. Technology and equipment for processing refractory, powder and composite materials[M]. Moscow: Metallurgy, 1989.

[75] Sokolkin Y V, Votinov A M, Tashkinov A A, et al. Technology of designing carbon-carbon composites and structures[M]. Moscow: Science. Fizmat-lit, 1996.

[76] Belov A F, Benediktov G P, Viskov A S, et al. The structure and properties of aviation materials[M]. Moscow: Metallurgy, 1989.

[77] Naughton B. The use of composite materials in technology[S]. Moscow: Mir, 1978.

[78] Tuchinsky L I. Composite materials obtained by the impregnation method [M]. Moscow: Metallurgy, 1986.

[79] U. K. Ministry of Defence. Design and Airworthiness Requirements for Service Aircraft[J]. Aeroplanes, 1988(1).

[80] U. S. Department of Defense. Aircraft Structures; General Specifications forAFGS-87221A[S], 1990.

[81] Barvinok V A, Bogdanovich V I, Bordakov P A, et al. Units, assembly and testing processes in the production of aircraft [M]. Moscow: Mechanical Engineering, 1996.

[82] Ushakov A E, Akimenko A A. Influence of the manufacturing method on the static strength of three-layer honeycomb structures with CFRP skin KMU-4E[J]. Aviation Industry, 1989(7).

[83] Krysin V N. Laminated glued structures in aircraft construction[M]. Moscow: Mashinostroenie, 1980.

[84] Zakharov V A. Construction of assemblies and parts from composite materials [M]. Moscow: MAI, 1992.

[85] Vanin G A. On the destruction of fibrous media[J]. Problems of aviation science, 1984.

[86] Vorobieva G N. On the standardization of services[J]. Standards and quality, 1998(1).

[87] Weinberger R A, Somoroff A R, Riley B L. U. S. Navy Certification of Composite Wings for the F-18 and the Advanced Harrier Aircraft [S]. AGARD Report No. 660, 1977.

[88] Whitehead R S. Certification of Primary Composite Aircraft Structures[J].
14th Symposium of the International Conference on Aeronautical Fatigue
(ICAF), 1987.

[89] Whitehead R S, Kan H P, Cordero R, et al. Certification Testing
Methodology for Composite Structure: Vol. II : Methodology Development
[S]. NADC-87042-60, Naval Air Development Center, 1986.

[90] Y₁ll₁kci Y K, Fındık F. A Survey of Aircraft Materials: Design for
Airworthiness and Sustainability[J]. Periodicals of engineering and natural
sciences, 2013,1(1).

[91] Zaitsev G P, Pashkov V A, et al. Strength and bone of unidirectional
organoplastic of the SVM type during tensile strength and durability of
elements of aircraft structures[J]. Interuniversity. Sat. scientific, 1979.

[92] Zakharov N G. Influence of ultrasonic treatment on the properties of the
adhesive composition [J]. Problems of improving the repair of aviation
equipment: Sat. scientific, 1989.

[93] Zholudev D A, Romanova V S, Semenov B I. Cast aluminum foamed
composites —a unique class of new materials for widespread use [J].
Tekhnologiya metallov,1999(11).